Frontiers in Applied Dynamical Systems: Reviews and Tutorials

Volume 10

Frontiers in Applied Dynamical Systems covers emerging topics and significant developments in the field of Dynamical Systems. It is an annual collection of invited review articles by leading researchers in dynamical systems and related areas. Contributions in this series should be seen as a portal for a broad audience of researchers in dynamical systems at all levels and can serve as advanced teaching aids. Each contribution provides an informal outline of a specific area, an interesting application, a recent technique, or a "how-to" for analytic methods and for computational algorithms, and a list of key references. All articles will be refereed.

Vasso Anagnostopoulou • Christian Pötzsche •
Martin Rasmussen

Nonautonomous Bifurcation Theory

Concepts and Tools

 Springer

Vasso Anagnostopoulou
Department of Primary Care and Public
Health
Brighton & Sussex Medical School
University of Sussex
Brighton, UK

Christian Pötzsche
Institut für Mathematik
University of Klagenfurt
Klagenfurt, Austria

Martin Rasmussen
Department of Mathematics
Imperial College London
London, UK

ISSN 2364-4532   ISSN 2364-4931   (electronic)
Frontiers in Applied Dynamical Systems: Reviews and Tutorials
ISBN 978-3-031-29841-7   ISBN 978-3-031-29842-4   (eBook)
https://doi.org/10.1007/978-3-031-29842-4

Mathematics Subject Classification: 37B55, 39A28, 34C23, 34D05, 37D10, 37G35

This Springer imprint is published by the registered company Springer Nature Switzerland AG
The registered company address is: Gewerbestrasse 11, 6330 Cham, Switzerland

Preface

The last 20 years have experienced an increased interest in evolutionary differential and difference equations subject to time-varying forcing or perturbation. Such studies require an extension of the classical theory which lead to the recent field of nonautonomous dynamical systems. Bifurcation theory, or the question for qualitative changes in the behaviour of time-variant equations under parameter variation, is a natural part of this theory. However, in contrast to autonomous problems, until now it is not clear what a nonautonomous bifurcation actually is, and so far, various different approaches to describe qualitative changes were suggested in the literature and denoted as 'bifurcation'. The aim of this book is to review, discuss and compare several concepts, to formulate them in a unified notation and to illustrate them by means of amenable examples. We additionally present certain relevant tools needed in a corresponding analysis. In short, we attempt to provide a concise survey of the area and equip the reader with suitable tools to tackle nonautonomous problems.

Nonautonomous bifurcation theory has various facets. On the one hand, the reader will discover natural generalisations of known patterns: under time-variant forcing, equilibria turn into periodic, quasi-periodic or bounded entire solutions as bifurcating objects, or, more generally, become copies of the base, in settings where the base set (that reflect the particular temporal forcing) is topologically well-behaved. On the other hand, in a more general nonautonomous setting, certain types of solutions such as equilibria, quasi-periodic or almost periodic solutions are no longer special and do not serve as bifurcating objects. As a consequence, appropriate other objects such as attractors that bifurcate need to be identified.

We will demonstrate in this book that nonautonomous dynamical systems give rise to several new, interesting and intrinsically nonautonomous phenomena. For instance, the imaginary axis (in continuous time) or the unit circle (in discrete time), as the boundary for the stability region, are in some sense 'wider' for nonautonomous equations. There are different possibilities to capture a hyperbolic setting, namely via the dichotomy spectrum or using Lyapunov exponents. Since the dichotomy spectrum consists of intervals, a loss of hyperbolicity happens in two steps, namely when a spectral interval enters, contains and when it leaves the stability boundary

0 (continuous time) or 1 (discrete time). During the nonhyperbolic regime, there might be a family of bounded entire solutions, but also chaotic behaviour can occur. Conversely, a hyperbolic situation can suddenly arise when a spectral interval causing nonhyperbolicity collapses into its boundary points.

The potential reader should be familiar with some basic knowledge on ordinary differential equations, (autonomous) dynamical systems and classical bifurcation theory, but beyond that we were aiming at a self-contained presentation. The contents of this text are of relevance for applied mathematicians, physicists and engineers. It might also serve as an addition to a modern course on dynamical systems on a MSc or PhD level. Nevertheless, providing detailed proofs throughout would have gone beyond the scope (and capacity) of this book.

The text starts with the fundamental concepts on nonautonomous dynamics and illustrative examples for bifurcations in time-variant equations. Our presentation covers finite-dimensional systems in continuous time (Chaps. 2–4 on differential equations are Part I), as well as in discrete time (Chaps. 5–7 on difference equations form Part II). For the sake of an efficient presentation, each part begins with standing assumptions and notes on the used terminology. Although the corresponding theories are parallel to a certain extent, we largely avoid redundancies and focus on features specific for the particular time line $\mathbb{R}$ or $\mathbb{Z}$. Many results are accompanied by examples, which also demonstrate different approaches to nonautonomous bifurcations and their merits in concrete situations. Finally, every section is concluded with additional remarks, further perspectives, hints to the related literature and possible research directions. An appendix on classical stability theory and Bohl/Lyapunov exponents closes these notes.

Acknowledgements

We are grateful to Álvaro Castañeda and Gonzalo Robledo (Santiago de Chile), Robert Skiba (Toruń) for hospitality, Peter E. Kloeden (Tübingen) for continued support, motivation and feedback, and Luca Arcidiacono (Munich), István Balázs, Ábel Garab (Szeged) for feedback on the manuscript. We thank the four anonymous referees for their helpful and constructive remarks and suggestions. Finally, we are grateful to Thomas Wanner (Fairfax VA) for his excellent handling of our manuscript.

Brighton, UK   Vasso Anagnostopoulou
Klagenfurt, Austria   Christian Pötzsche
London, UK   Martin Rasmussen
November, 2022

Contents

Notation

Throughout these notes, the *time line* is denoted by $\mathbb{T}$ and stands either for the real numbers $\mathbb{R}$ (continuous time) or the integers $\mathbb{Z}$ (discrete time). We write $\mathbb{T}_0^+$ and $\mathbb{T}_0^-$ for the *half lines* of nonnegative respectively nonpositive elements of $\mathbb{T}$.

In a metric space (X, d), the *distance* of a point $x \in X$ from a subset $A \subseteq X$ is $\operatorname{dist}(x, A) := \inf_{a \in A} d(x, a)$ and we write

$$d(A, B) := \sup_{a \in A} \inf_{b \in B} d(a, b), \qquad h(A, B) := \max\{d(A, B), d(B, A)\}$$

for the *Hausdorff semi-distance* respectively the *Hausdorff distance* of $A, B \subseteq X$. The complete σ-algebra containing the Borel sets of X is abbreviated as $\mathfrak{B}(X)$.

Let X, Y be normed spaces and $\rho > 0$. We write $B_\rho(x)$, $\bar{B}_\rho(x)$ for the open respectively closed ρ-neighbourhood of a point $x \in X$. Furthermore, $GL(X)$ are the bounded invertible operators on X and $L_k(X, Y)$, $k \in \mathbb{N}_0$, are the bounded, symmetric, k-linear maps from the k-fold Cartesian product X^k to Y, supplemented by $L_0(X, Y) := Y$. The space of bounded continuous functions $\phi : \mathbb{R} \to X$ is $BC(X)$ and $\ell^\infty(X)$ stands for the bounded sequences $\phi : \mathbb{Z} \to X$; both spaces are equipped with the sup-norm.

On the d-dimensional space $\mathbb{R}^d$, we use $\langle x, y \rangle := \sum_{k=1}^d x_k y_k$ as inner product and write $\|\cdot\|$ for the induced norm. The set of all compact subsets of $\mathbb{R}^d$ is $\mathfrak{K}(\mathbb{R}^d)$.

The normed space of all real matrices with d rows and m columns is abbreviated as $\mathbb{R}^{d \times m}$. Given such a matrix $A \in \mathbb{R}^{d \times m}$, then $A^T \in \mathbb{R}^{m \times d}$ is its transpose and $A^\dagger \in \mathbb{R}^{m \times d}$ the pseudo-inverse. Moreover, $N(A) \subseteq \mathbb{R}^m$ stands for the kernel (null space) and $R(A) \subseteq \mathbb{R}^d$ for the range. We write $\operatorname{tr} B$, $\det B$ for the trace respectively the determinant of a square matrix $B \in \mathbb{R}^{d \times d}$ and id for the identity matrix. The general linear group $GL(\mathbb{R}^d) \subseteq \mathbb{R}^{d \times d}$ consists of invertible matrices and the special linear group $SL(\mathbb{R}^d) \subseteq \mathbb{R}^{d \times d}$ are the matrices having determinant 1.

Subsets of $\mathbb{T} \times \mathbb{R}^d$ are called *nonautonomous sets* and denoted by scripted letters $\mathcal{A}, \mathcal{B}, \ldots$. The t-*fibre* of a nonautonomous set $\mathcal{A}$ is $\mathcal{A}(t) := \{x \in \mathbb{R}^d : (t, x) \in \mathcal{A}\}$

for $t \in \mathbb{T}$. Moreover, we set $\mathcal{B}_\rho(\mathcal{A}) := \{(t,x) \in \mathbb{T} \times \mathbb{R}^d : \text{dist}(x,\mathcal{A}(t)) < \rho\}$ for the ρ-neighbourhood of a nonautonomous set $\mathcal{A}$. Finally, when dealing with skew-product flows over a base space Ω, a corresponding terminology is also used for subsets $\mathcal{A}$ of $\Omega \times \mathbb{R}^d$ rather than $\mathbb{T} \times \mathbb{R}^d$.

Chapter 1
Introduction

The term *bifurcation* was introduced by Henri Poincaré in 1885 in the first mathematical paper [184] exhibiting such a behaviour. As indicated by the name, bifurcation (from the Latin *bifurcātus*: forked in two) theory consists of (at least) two branches:

(1) In nonlinear analysis it is also called *branching theory* and deals with changes in the solution structure to abstract (nonlinear) equations under parameter variation [48, 60, 132, 229]. This field can be traced back at least to Euler (1707–1783) and his studies on the critical load which a column (a rod) can bear. Milestones were the Lyapunov–Schmidt reduction and the Crandall–Rabinowitz results on bifurcations from a given branch. Yet, in this text the problems are evolutionary differential or difference equations.

(2) In dynamical systems the theory becomes richer since a bifurcation typically goes hand in hand with a change of stability properties to particular reference solutions [57, 94, 148, 228]. Here, the Hopf bifurcation and the centre manifold theory must be mentioned as cornerstones.

Classical dynamical bifurcation theory focuses on autonomous, parametrised differential and difference equations of the form

$$\dot{x} = g(x, \lambda) \quad \text{or} \quad x_{t+1} = g(x_t, \lambda) \tag{1.1}$$

with a sufficiently smooth right hand side $g : \mathbb{R}^d \times \Lambda \to \mathbb{R}^d$, involving a *bifurcation parameter* λ. We assume that the *parameter space* Λ is a metric space, typically given by an open subset of $\mathbb{R}^n$ or some Banach space. A central bifurcation-theoretical question is how stability, attractiveness, and multiplicity properties of invariant sets for (1.1) change under variation of λ? In the simplest situation, these invariant sets are given by equilibria (fixed points) or periodic solutions of (1.1).

For some fixed parameter value $\lambda^* \in \Lambda$, an equilibrium x^* of (1.1) is called *hyperbolic* if the partial derivative $D_1 g(x^*, \lambda^*) \in \mathbb{R}^{d \times d}$ possesses no eigenvalue on the stability boundary; in continuous time, this is the imaginary axis, whereas the complex unit circle $\mathbb{S}^1$ decides on stability in discrete time. The implicit function

© The Author(s), under exclusive license to Springer Nature Switzerland AG 2023
V. Anagnostopoulou et al., *Nonautonomous Bifurcation Theory*, Frontiers
in Applied Dynamical Systems: Reviews and Tutorials 10,
https://doi.org/10.1007/978-3-031-29842-4_1

theorem allows a unique continuation $g(x(\lambda), \lambda) \equiv 0$ or $g(x(\lambda), \lambda) \equiv x(\lambda)$ of x^* in a neighbourhood of λ^*. In particular, hyperbolicity rules out bifurcations understood as topological changes in the set of equilibria $\{x \in \mathbb{R}^d : g(x, \lambda) = 0\}$ or fixed-points $\{x \in \mathbb{R}^d : g(x, \lambda) = x\}$, respectively, near a reference pair (x^*, λ^*) or a stability change of x^*. Indeed, according to [148, p. 57, Definition 2.11]:

> *The appearance of a topologically nonequivalent phase portrait*
> *under variation of parameters is called a bifurcation.*

On the one hand, autonomous bifurcation theory is well-developed. Eigenvalues on the stability boundary at a *critical parameter* value λ^* give rise to various classical bifurcation scenarios. Such prototypical examples include fold (saddle-node), transcritical or pitchfork bifurcations (eigenvalue 0 or 1, respectively), period-doubling bifurcations in discrete time (eigenvalue -1), and Hopf or Sacker–Neimark bifurcations (a pair of complex conjugate eigenvalues for $d \geq 2$). For one-dimensional parameter spaces Λ one can distinguish between *super-* and *subcritical bifurcations*, depending whether the bifurcating object occurs for $\lambda > \lambda^*$ or $\lambda < \lambda^*$. Via centre manifold theory or Lyapunov–Schmidt reduction, higher-dimensional problems can be reduced to the above situations. Moreover, normal form theory allows a classification of bifurcation scenarios by finding an algebraically most simple representation of (1.1). It can be said that the dynamical bifurcation theory for autonomous equations (1.1) has reached a remarkable maturity concerning analytical as well as numerical aspects and various effective computational tools are available [70, 92, 143].

On the other hand, for nonautonomous equations, it is not even quite clear what a bifurcation actually is, since it is problematic to obtain an ambient notion of topological equivalence; we elaborate on that in Sect. 1.3. As a first step one tried to mimic the autonomous situation and to come up with nonautonomous counterparts to the classical bifurcation patterns. Beyond that, also new phenomena arose, like the shovel bifurcation [189] or the two-step scenario [121].

This text deals with explicitly time-variant problems: After the pioneering work of, e.g. R.J. Sacker and G.R. Sell in the 1970s, about 20 years ago a new interest in nonautonomous dynamics began culminating in monographs like [43, 139, 186] or [36, 141] having a focus on attractors. In this context, the question for a bifurcation theory arose quite naturally, where it is common sense not to understand periodic equations as being nonautonomous. This is due to the fact that tools like Floquet theory [96, pp. 117ff, III.7], period or Poincaré maps essentially allow to reduce them to time-independent problems.

There are various reasons to study systems under time-varying influences. Indeed, even in the time-invariant setting of autonomous equations

$$\dot{x} = g(x) \quad \text{or} \quad x_{t+1} = g(x_t), \tag{1.2}$$

one often encounters intrinsically nonautonomous problems, to which neither the classical autonomous theory as presented in, for instance, [57, 94, 112, 148, 228], nor the numerical routines in, for example, [70, 92, 143], can be applied. We now describe initially autonomous problems related to (1.2) leading to nonautonomous differential or difference equations of the form (1.4) below.

- The investigation of the dynamical behaviour of (1.2) along an entire reference solution $\phi : \mathbb{R} \to X$, respectively, $\phi : \mathbb{Z} \to X$ that is not constant or periodic leads to the (nonautonomous) *equation of perturbed motion*

$$\dot{x} = g(x + \phi(t)) - g(\phi(t)) \quad \text{or} \quad x_{t+1} = g(x_t + \phi(t)) - g(\phi(t)), \quad (1.3)$$

which has the trivial solution (representing the reference solution ϕ). Note that (1.3) is of the form (1.4) with $f(t, x) = g(x + \phi(t)) - g(\phi(t))$.
- Replacing the constant parameter λ by a time-dependent function $\lambda(t) \in \Lambda$ in (1.1) results in a *parametrically perturbed* equation of the form

$$\dot{x} = g(x, \lambda(t)) \quad \text{or} \quad x_{t+1} = g(x_t, \lambda(t)),$$

which is nonautonomous of the form (1.4) with $f(t, x) = g(x, \lambda(t))$. This situation is highly relevant for applications, in order to mimic control or regulation strategies via the function λ. One might think of quasi-periodic driving in geophysical applications (gravitational forces of planets) or time-varying growth rates and carrying capacities in ecological systems.
- Numerical discretisations of ordinary differential equations $\dot{x} = g(x)$ with adaptive time-steps $h_t > 0$ yield nonautonomous difference equations. In the simplest case of the forward Euler method, they read as

$$x_{t+1} = x_t + h_t g(x_t)$$

which is nonautonomous of the form (1.4) with $f(t, x) = x + h_t g(x)$.

Given this, nonautonomous bifurcation theory explores essentially different approaches. They depend on both the spectral theory of the underlying linearisations (cf. the dichotomy vs. the Lyapunov spectrum in Chaps. 2 and 5), as well as the 'canonical' space in which bifurcating objects (attractors, solutions, minimal sets) are sought. The required tools come from stability and dynamical systems theory, as well as from nonlinear functional analysis.

1.1 Nonautonomous Dynamical Systems

In some parts of this work, we aim at treating continuous and discrete time simultaneously. We use the *time line* $\mathbb{T} = \mathbb{R}$ when dealing with differential equations, and we use $\mathbb{T} = \mathbb{Z}$ when being concerned with difference equations. *Half lines* are abbreviated as $\mathbb{T}_\tau^+ := \{t \in \mathbb{T} : t \geq \tau\}$ and $\mathbb{T}_\tau^- := \{t \in \mathbb{T} : t \leq \tau\}$ for some $\tau \in \mathbb{T}$.

First, we quickly discuss some notions from the classical theory of autonomous dynamical systems. Under appropriate conditions for global existence of solutions, the solutions of an autonomous differential or difference equation of the form

$$\dot{x} = f(x) \quad \text{or} \quad x_{t+1} = f(x_t)$$

give rise to a *dynamical system* (or *flow*) on a *state space* Ω (usually a metric space), that is a continuous mapping $\theta : \mathbb{T} \times \Omega \to \Omega$ satisfying

$$\theta(0, \omega) = \omega \qquad\qquad \text{for all } \omega \in \Omega,$$
$$\theta(t + s, \omega) = \theta(t, \theta(s, \omega)) \qquad\qquad \text{for all } s, t \in \mathbb{T} \text{ and } \omega \in \Omega.$$

Since the equation is autonomous (i.e. the right hand side does not depend explicitly on time), the evolution of the system only depends on the initial value and the elapsed time, but not the initial time. Therefore, it is sufficient to illustrate the behaviour of autonomous equations in the state space, which consists of orbits.

Here, the set $\theta(\mathbb{T}, \omega) \subseteq \Omega$ is denoted as *orbit* through a point $\omega \in \Omega$ and

$$L^+(\omega) := \bigcap_{T \geq 0} \overline{\bigcup_{t \geq T} \{\theta(t, \omega)\}}$$

as *forward limit set* of $\omega \in \Omega$. A subset $A \subseteq \Omega$ is called *invariant* (or *θ-invariant* to emphasise the particular dynamical system), if $\theta(t, A) = A$ holds for every $t \in \mathbb{T}$. Such a nonempty and compact set $A \subseteq \Omega$ is *minimal*, if A contains no nonempty, closed and invariant proper subset, and a flow is called *minimal*, if the state space Ω itself is minimal. One can show that θ is minimal if and only if every orbit $\theta(\mathbb{T}, \omega)$, $\omega \in \Omega$, is dense in Ω [169, p. 375].

Example 1.1.1 (Kronecker Flow) *Suppose* $\Omega := \mathbb{R}^n / \mathbb{Z}^n$ *is the standard n-torus. For frequencies* $\varrho_1, \ldots, \varrho_n \in \mathbb{R}$ *a* Kronecker flow $\theta : \mathbb{T} \times \Omega \to \Omega$ *is*

$$\theta(t, \omega) := (\omega_1 + \varrho_1 t, \ldots, \omega_n + \varrho_n t) \quad (\text{mod } 1).$$

If the reals $\varrho_1, \ldots, \varrho_n$ *are* rationally independent, *i.e.* $\sum_{j=1}^n q_j \varrho_j = 0$ *implies that* $q_1 = \ldots = q_n = 0$ *for all* $q_1, \ldots, q_n \in \mathbb{Q}$, *then the flow* θ *is minimal, because every orbit* $\theta(\mathbb{T}, \xi)$, $\xi \in \Omega$, *must be dense in* Ω *[129, p. 29, Proposition 1.4.1].*

A flow $\theta : \mathbb{T} \times \Omega \to \Omega$ on a compact metric space Ω is amenable to a statistical analysis due to the existence of invariant measures [169]. A probability measure $\mu : \mathfrak{B}(\Omega) \to [0, 1]$, where $\mathfrak{B}(\Omega)$ is a complete σ-algebra containing the Borel sets of Ω, is called *invariant* if $\mu(B) = \mu(\theta(1, B))$ for all $B \in \mathfrak{B}$. In this situation, the flow θ is also called *measure-preserving*. In addition, θ is called *ergodic*, if the θ-invariant subsets of Ω have measure 0 or 1, and *uniquely ergodic*, if μ is the unique invariant probability measure having this property. We note that minimal and uniquely ergodic dynamical systems are also called *strictly ergodic*, and this class includes also the above Kronecker flow.

In contrast to autonomous differential and difference equations, solutions of a nonautonomous differential or difference equation of the form

$$\dot{x} = f(t, x) \quad \text{or} \quad x_{t+1} = f(t, x_t) \tag{1.4}$$

cannot be described by flow θ, and a corresponding notion of a *nonautonomous dynamical system* needs an additional parameter indicating the initial time. In the following, we present two versions of nonautonomous dynamical systems (given by *processes* and *skew product flows*) that involve this additional parameter. For an in-depth introduction to nonautonomous dynamical systems, see [43, 139, 141, 186].

Before continuing, we would like to alert the reader: many people suggest a transformation to write nonautonomous equations (1.4) as autonomous ones

$$\begin{cases} \dot{t} = 1, \\ \dot{x} = f(t, x), \end{cases} \qquad \begin{cases} u_{t+1} = u_t + 1, \\ x_{t+1} = f(u_t, x_t), \end{cases} \tag{1.5}$$

by considering time as an additional variable. This approach does not offer any simplifications. First, (1.5) do not have equilibria, all solutions are unbounded and therefore every limit set is empty, and attractors do not exist. Second, in the continuous-time case, the dependence of f on t has to be as smooth as in x, which rules out various interesting applications like discontinuous time-dependence or even Carathéodory equations [146] being central in control theory [52].

1.1.1 Processes

Assume that a nonautonomous differential or difference equation (1.4) satisfies conditions guaranteeing existence and uniqueness of solutions forward in time (if $\mathbb{T} = \mathbb{R}$). Then the general solution (in forward time) of the system (1.4) is given by a continuous function $\varphi : \{(t, \tau) \in \mathbb{T}^2 : t \geq \tau\} \times X \to X$, where

$$\begin{aligned} \varphi(\tau, \tau, \xi) &= \xi \quad \text{for all } \tau \in \mathbb{T} \text{ and } \xi \in X, \\ \varphi(t, \tau, \xi) &= \varphi(t, s, \varphi(s, \tau, \xi)) \quad \text{for all } t \geq s \geq \tau \text{ and } \xi \in X, \end{aligned} \tag{1.6}$$

and X is the *state space* of (1.4). We refer to an abstract function φ with the above properties as a *process* (sometimes also called 2-*parameter semigroup*), and we typically require that the state space X is metric. One speaks of an *invertible process*, if φ can be extended to $\mathbb{T} \times \mathbb{T} \times X$ preserving the above properties.

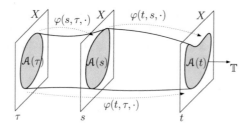

Figure 1.1 Process φ on the extended state space $\mathbb{T} \times X$ and an invariant nonautonomous set $\mathcal{A} \subset \mathbb{T} \times X$

In contrast to the case of autonomous dynamical systems, where the interesting objects (such as fixed points, periodic orbits, attractors, or invariant manifolds) are subsets of the state space X, for nonautonomous dynamical systems, these objects are subsets of the *extended state space* $\mathbb{T} \times X$. In particular, rather than with orbits, one works with solution curves.

Any subset $\mathcal{A} \subseteq \mathbb{T} \times X$ of the extended state space is called a *nonautonomous set*, and for each time $t \in \mathbb{T}$, the *t-fibre* of a nonautonomous set $\mathcal{A}$ is given by

$$\mathcal{A}(t) := \{x \in X : (t, x) \in \mathcal{A}\} \ .$$

For a *compact* nonautonomous set $\mathcal{A}$ all the fibres $\mathcal{A}(t)$, $t \in \mathbb{T}$, are compact in X. We denote a nonautonomous set $\mathcal{A}$ as *invariant* with respect to a process φ if one has $\mathcal{A}(t) = \varphi(t, \tau, \mathcal{A}(\tau))$ for $t \geq \tau$ (see Fig. 1.1). The simplest form of an invariant nonautonomous set can be identified with an *entire solution*, which is determined by a function $\phi : \mathbb{T} \to X$ so that $\varphi(t, \tau, \phi(\tau)) = \phi(t)$. The associated nonautonomous set $\{(t, \phi(t)) : t \in \mathbb{T}\}$ is given by its *solution curve* and has singleton fibres.

1.1.2 Skew Product Flows

The above introduced process formulation of a nonautonomous dynamical system is very intuitive and applies to a large class of problems, but it lacks topological structure that can simplify the analysis and better illustrate the behaviour of specific non-autonomous problems. In this subsection, we study its alternative, so-called skew product flows. In contrast to processes, the dependence on the initial time τ is replaced by a dependence on an element of the state space of a dynamical system that models the time-dependent driving of the skew product flow. This allows to capture very different types of time-dependence and to recover them in bifurcating objects.

Let the continuous function $\theta : \mathbb{T} \times \Omega \to \Omega$ be a dynamical system on a metric space Ω. For a fixed $\omega \in \Omega$, let

$$\dot{x} = f(\theta(t, \omega), x) \quad \text{or} \quad x_{t+1} = f(\theta(t, \omega), x_t) \tag{1.7}$$

denote a nonautonomous differential or difference equation. We assume global existence and uniqueness of solutions in forward time of (1.7) (which is only an issue in continuous time), and we denote the solution of (1.7) satisfying the initial condition $x(0) = \xi \in X$ by $\varphi(\cdot, \omega, \xi)$. It follows that φ satisfies

$$\varphi(0, \omega, \xi) = \xi \quad \text{for all } \omega \in \Omega \text{ and } \xi \in X \ ,$$
$$\varphi(t + s, \omega, \xi) = \varphi(t, \theta(s, \omega), \varphi(s, \omega, \xi)) \quad \text{for all } t, s \geq 0, \omega \in \Omega \text{ and } \xi \in X \ .$$

The pair (θ, φ), given by a dynamical system $\theta : \mathbb{T} \times \Omega \to \Omega$ and a continuous function φ satisfying the above two identities, is called a *skew product flow*. Now θ is denoted as *base flow*, and we refer to Ω as the *base set*, while φ is named *cocycle* (see Fig. 1.2).

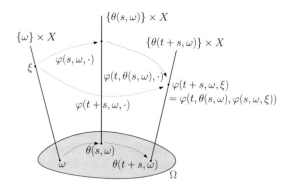

Figure 1.2 Skew product flow (θ, φ) on the extended state space $\Omega \times X$ driven by the base flow $\theta : \mathbb{T} \times \Omega \to \Omega$ over the base set Ω

We note that any skew product flow (θ, φ) defines an autonomous dynamical system on the *extended state space* $\Omega \times X$ via

$$(\theta, \varphi)(t, \omega, \xi) = (\theta(t, \omega), \varphi(t, \omega, \xi)). \tag{1.8}$$

Analogously to processes, any subset $\mathcal{A} \subseteq \Omega \times X$ of the extended state space is called a *nonautonomous set*, and for each $\omega \in \Omega$, the ω-*fibre* of a nonautonomous set $\mathcal{A}$ is defined as

$$\mathcal{A}(\omega) := \{x \in X : (\omega, x) \in \mathcal{A}\}.$$

One denotes $\mathcal{A}$ as *compact*, if the fibres $\mathcal{A}(\omega)$ are compact for each $\omega \in \Omega$. We call a nonautonomous set $\mathcal{A}$ *invariant* with respect to a skew product flow (θ, φ) if $\mathcal{A}(\theta(t, \omega)) = \varphi(t, \omega, \mathcal{A}(\omega))$ for all $t \geq 0$. The analogue of an entire solution in the context of skew product flows is an *invariant graph*, i.e. a function $\phi : \Omega \to X$ such that $\varphi(t, \omega, \phi(\omega)) = \phi(\theta(t, \omega))$ holds. Then the associated nonautonomous set $\{(\omega, \phi(\omega)) : \omega \in \Omega\}$ is given by its graph and possesses singleton fibres. Finally, a nonautonomous set $\mathcal{M} \subseteq \Omega \times X$ is called a *copy of the base*, provided every fibre $\mathcal{M}(\omega) \subset X$ is a singleton.

Any process $\varphi : \{(t, \tau) \in \mathbb{T}^2 : t \geq \tau\} \times X \to X$ can be written as a skew product flow $\hat{\varphi}$, where $\Omega = \mathbb{T}$, $\theta(t, \omega) = \omega + t$ for $t \in \mathbb{T}$ and $\omega \in \Omega$, and $\hat{\varphi}(t, \omega, \xi) = \varphi(t + \omega, \omega, \xi)$ for all $t \geq 0$, $\omega \in \Omega$ and $\xi \in X$. Note that the first argument of $\hat{\varphi}$ denotes elapsed time after $\omega \in \Omega$, while the first argument of φ denotes the current time. Such an embedding of processes into skew product flows does not have any advantage, since the resulting base set Ω is a non-compact metric space. Whenever considering skew product flows, in particular, we will require compact spaces Ω, which will facilitate the analysis, since in many examples, invariant nonautonomous sets $\mathcal{A}$ will be compact subsets of the extended state space $\Omega \times X$, and this can only happen when Ω is compact. We note that many important classes of nonautonomous differential or difference equations of the form (1.4) (such as quasi or almost periodic equations) lead to compact base sets Ω [209, 213]. This is not true for 'autonomised' equations (1.5), whose base sets are $\mathbb{R}$, respectively, $\mathbb{Z}$. This provides another reason why making an equation autonomous as in (1.5) is not useful.

1.2 Examples of Nonautonomous Bifurcations

After these preliminaries on the process and skew product formulation of nonautonomous dynamical system, we study concrete differential or difference equations

$$\dot{x} = f(t, x, \lambda) \quad \text{or} \quad x_{t+1} = f(t, x_t, \lambda), \tag{1.9}$$

depending on a parameter $\lambda \in \mathbb{R}$. By means of simple examples we aim to demonstrate certain (possibly characteristic) features of nonautonomous bifurcations. The
process generated by (1.9) is denoted by φ_λ in order to indicate the λ-dependence.

We first observe that nonautonomous differential or difference equations usually
do not have equilibria or periodic solutions. This means that typically, we do not
have $0 = f(t, x, \lambda)$ for some $x \in X$ and all $t \in \mathbb{R}$ in case of a differential equation,
or $x = f(t, x, \lambda)$ for some $x \in X$ and every $t \in \mathbb{Z}$ in case of a difference equation.
This gives rise to the following question:

> *If there are no equilibria or periodic solutions,*
> *what objects do bifurcate in nonautonomous dynamical systems?*

An adequate answer to this question forces us to enlarge the set of objects in which
we look for bifurcating objects. For instance, many nonautonomous bifurcations
involve qualitative changes in the attractors of the system. A nonempty, compact,
and invariant nonautonomous set $\mathcal{A}$ of a process φ, defined on a metric space (X, d),
is called a *global pullback attractor* if for any nonempty bounded set $B \subset X$, we
have

$$\lim_{\tau \to -\infty} d(\varphi(t, \tau, B), \mathcal{A}(t)) = 0 \quad \text{for all } t \in \mathbb{T},$$

where $d(A, B) := \sup_{a \in A} \inf_{b \in B} d(a, b)$ is the *Hausdorff semi-distance* of two
subsets A and B of X.

Before studying bifurcations involving attractors, we first consider a situation
where globally attractive equilibria of an autonomous systems persist as global pullback attractors under a nonautonomous perturbation.

Example 1.2.1 *Let* $b : \mathbb{R} \to \mathbb{R}$ *be a bounded and continuous function, and consider
the scalar linear nonautonomous differential equation*

$$\dot{x} = -x + \lambda b(t). \tag{1.10}$$

For $\lambda = 0$, *it is autonomous and has a unique fixed point* $x = 0$. *This equilibrium
is globally attracting and the unique bounded and entire solution of* (1.10). *In the
nonautonomous case* $\lambda \neq 0$, *the zero of the right hand side of* (1.10), *given by*
$x(t) = \lambda b(t)$, *does not have a dynamical meaning for two different reasons. First,
being merely a continuous function, on a purely formal basis,* $x(t) = \lambda b(t)$ *cannot
be a solution to* (1.10). *Secondly, even if* b *is assumed to be differentiable, then the
solution identity* $\lambda \dot{b}(t) \equiv \dot{x}(t) \equiv -x(t) + \lambda b(t) \equiv 0$ *on* $\mathbb{R}$ *requires* $\lambda = 0$ *or a
constant function* b, *which both refers to the situation that* (1.10) *is autonomous.*

Yet, the differential equation (1.10) *still admits a unique bounded entire solution*

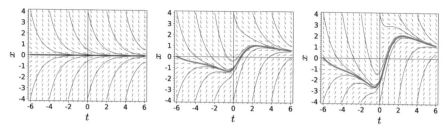

Figure 1.3 Persistence of bounded entire solutions: Solution curves (blue) of the linear differential equation (1.10) with inhomogeneity $b(t) = \frac{t}{2+t^2}$ and $\lambda = 0$ (left), $\lambda = 3$ (centre), $\lambda = 6$ (right), and the unique bounded solution ϕ_λ (green)

$$\phi_\lambda(t) := \lambda \int_{-\infty}^{t} e^{s-t} b(s)\, ds \quad \text{for all } t \in \mathbb{R}.$$

Hence, the equilibrium $\phi_0 \equiv 0$ of (1.10) for $\lambda = 0$ persists as a bounded entire solution ϕ_λ when $\lambda \neq 0$, which is globally attracting in the sense that the nonautonomous set corresponding to the entire solution ϕ_λ is a global pullback attractor (see Fig. 1.3).

We see later that certain bifurcations in nonautonomous dynamical systems can be observed if attractors change discontinuously under parameter variation. In this simple example, the global attractor is given by an entire solution ϕ_λ for any $\lambda \in \mathbb{R}$, and we do not observe such a bifurcation here.

We note that it is easy to see that if the function b is periodic, then the entire solution ϕ_λ is periodic as well. It also follows that if b is quasi-periodic or almost periodic, then ϕ_λ is quasi-periodic or almost periodic, respectively (see, e.g. [96, p. 148, Theorem 1.1] or [79, p. 143, Theorem 8.1]).

This linear example leads to the question whether equilibria of the autonomous systems persist as bounded entire solutions under nonautonomous perturbations and whether this behaviour can also be observed for nonlinear equations. The results in Sects. 2.3 and 5.3 confirm that this is generically true in the sense that an equilibrium of an autonomous system has to be hyperbolic in order to persist under nonautonomous perturbations.

We now study a scenario where the hyperbolicity condition is violated, and where bifurcations in form of changes in the set of bounded solutions and their stability can be observed.

Example 1.2.2 *Consider the scalar linear nonautonomous difference equation*

$$x_{t+1} = \lambda x_t + \frac{1}{1+|t|}, \tag{1.11}$$

and let φ_λ denote the induced process. As the following cases show, the asymptotic behaviour of (1.11) crucially depends on the real parameter λ.

- *Case $|\lambda| < 1$. Similarly to Example 1.2.1, it follows that the difference equation (1.11) has a unique bounded entire solution $\phi_\lambda(t) := \sum_{s=-\infty}^{t-1} \frac{\lambda^{t-s-1}}{1+|s|}$ for* *all $t \in \mathbb{Z}$. This solution is uniformly asymptotically stable (see Appendix A.1), and the nonautonomous set $A_\lambda := \phi_\lambda$ is the global pullback attractor.*

- *Case $\lambda = 1$. It is straightforward to obtain an explicit representation of the process φ_1. It involves the harmonic series, is given by*

$$\varphi_1(t,\tau,\xi) = \begin{cases} \xi + \sum_{s=\tau}^{t-1} \frac{1}{1+|s|} & : \ t \geq \tau, \\ \xi - \sum_{s=t}^{\tau-1} \frac{1}{1+|s|} & : \ t < \tau, \end{cases}$$

and thus, there exist no bounded entire solutions to (1.11).

- *Case $\lambda = -1$. Analogously to the case $\lambda = 1$, the process φ_{-1} is given by*

$$\varphi_{-1}(t,\tau,\xi) = \begin{cases} (-1)^{t-\tau}\xi - \sum_{s=\tau}^{t-1} \frac{(-1)^{t-s}}{1+|s|} & : \ t \geq \tau, \\ (-1)^{t-\tau}\xi + \sum_{s=t}^{\tau-1} \frac{(-1)^{t-s}}{1+|s|} & : \ t < \tau, \end{cases}$$

and consequently every solution of (1.11) is bounded, since it involves the alternating harmonic series.

- *Case $|\lambda| > 1$. The difference equation (1.11) has a unique bounded entire solution $\phi_\lambda(t) := -\frac{1}{\lambda}\sum_{s=t}^{\infty} \frac{\lambda^{t-s}}{1+|s|}$, which has repelling properties.*

For the critical parameters $\lambda = \pm 1$, the linear difference equation (1.11) changes not only its stability. At $\lambda = -1$, the number of bounded entire solutions explodes, in contrast to uniqueness of bounded entire solutions for λ close to -1. In addition, near the parameter value $\lambda = 1$, there are unique bounded entire solutions, while there is none for $\lambda = 1$.

After these two linear examples, we study a nonlinear nonautonomous differential equation admitting a bifurcation of pitchfork type. It is a generalisation of the prototypical autonomous example (*Bernoulli equation*)

$$\dot{x} = \lambda x - \beta x^3$$

exhibiting a pitchfork bifurcation (here λ is the bifurcation parameter, and $\beta > 0$). This means that for $\lambda \leq 0$, there is only one equilibrium, given by $x = 0$, which is attractive. For $\lambda > 0$, however, the trivial equilibrium becomes repulsive, and two other attractive equilibria $x = \pm\sqrt{\lambda}$ bifurcate from it. In addition, the global attractor is given by $\{0\}$ for $\lambda \leq 0$ and by the interval $\left[-\sqrt{\frac{\lambda}{\beta}}, \sqrt{\frac{\lambda}{\beta}}\right]$ for $\lambda > 0$, which means that the autonomous pitchfork bifurcation can also be described by the trivial attractor becoming nontrivial at the bifurcation point.

Example 1.2.3 (Nonautonomous Pitchfork Bifurcation) *For a continuous function $b : \mathbb{R} \to (0,\infty)$, consider the nonautonomous Bernoulli equation*

$$\dot{x} = \lambda x - b(t)x^3, \tag{1.12}$$

which generates the process (see [150])

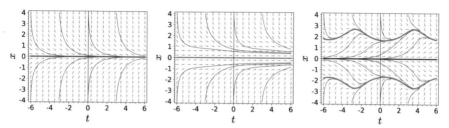

Figure 1.4 Pitchfork bifurcation from Example 1.2.3: Solution curves (blue) of the Bernoulli equation (1.12) with coefficient $b(t) = \frac{1}{4}(1 + \frac{1}{2}\cos t)$ and $\lambda = -1$ (left, exponential decay to 0), $\lambda = 0$ (centre, subexponential decay to 0), $\lambda = 1$ (right), and the attractive bounded solutions $\phi_\lambda^\pm$ (green)

$$\varphi_\lambda(t, \tau, \xi) = \frac{e^{\lambda(t-\tau)}\xi}{\sqrt{1 + 2\xi^2 \int_\tau^t e^{2\lambda(s-\tau)} b(s)\, ds}}.$$

If there exists a $c > 0$ such that

$$\frac{1}{c} \le b(t) \le c \quad \text{for all } t \in \mathbb{R}, \tag{1.13}$$

then analogously to the autonomous counterpart, (1.12) admits the trivial solution, which forms the global pullback attractor when $\lambda \le 0$, and which is repulsive for $\lambda > 0$. There are also analogues for the two attractive equilibria created at the autonomous pitchfork bifurcation point, given by two (locally) attractive solutions. They can be obtained via pullback limits $\tau \to -\infty$ for $\xi > 0$, respectively, $\xi < 0$:

$$\phi_\lambda^\pm(t) := \lim_{\tau \to -\infty} \varphi_\lambda(t, \tau, \xi) = \frac{\pm 1}{\sqrt{2 \int_{-\infty}^t e^{2\lambda(s-t)} b(s)\, ds}}.$$

It follows quickly from the explicit representation of φ_λ that the global pullback attractor $\mathcal{A}_\lambda$ for $\lambda > 0$ has the fibres $\mathcal{A}_\lambda(t) := [\phi_\lambda^-(t), \phi_\lambda^+(t)]$, where $t \in \mathbb{R}$. Since the global pullback attractor for $\lambda \le 0$ is given by $\mathcal{A}_\lambda = \mathbb{R} \times \{0\}$, the nonautonomous pitchfork bifurcation can be understood as attractor bifurcation as well. In Fig. 1.4, the function b is 2π-periodic and thus also the entire solutions $\phi_\lambda^\pm$ and the pullback attractor $\mathcal{A}_\lambda$ share this property.

While the above nonautonomous examples generalise bifurcations found in autonomous systems, the following scenario is intrinsically nonautonomous. The idea is to generate a nonhyperbolic situation by concatenating two hyperbolic systems.

Example 1.2.4 (Shovel Bifurcation) *Let $\lambda > 0$, and consider the scalar linear difference equation*

$$x_{t+1} = a_t(\lambda)x_t, \qquad \text{where } a_t(\lambda) := \begin{cases} \frac{1}{2} + \lambda & : \quad t < 0, \\ \lambda & : \quad t \ge 0. \end{cases} \tag{1.14}$$

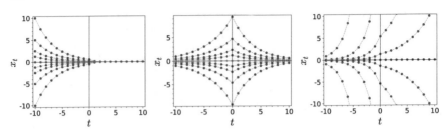

Figure 1.5 Shovel bifurcation from Example 1.2.4: Solution sequences (dotted) of the linear difference equation (1.14) with $\lambda = \frac{1}{4}$ (left, unique bounded entire solution), $\lambda = \frac{3}{4}$ (centre, every solution is bounded), $\lambda = \frac{5}{4}$ (right, unique bounded entire solution)

We distinguish three different parameter constellations:

- *Case $\lambda \in (0, \frac{1}{2})$. The unique bounded entire solution is trivial and uniformly asymptotically stable. The global attractor of (1.14) is given by $\mathcal{A}_\lambda = \mathbb{Z} \times \{0\}$ (see Fig. 1.5, left).*
- *Case $\lambda \in (\frac{1}{2}, 1)$. For this parameter regime, every solution is bounded. Moreover, (1.14) is exponentially stable, but not uniformly asymptotically stable on the whole time line $\mathbb{Z}$. There exists no global attractor (see Fig. 1.5, centre).*
- *Case $\lambda > 1$. The unique bounded entire solution is the trivial solution, which is repulsive, and there is no global attractor (see Fig. 1.5, right).*

This means that the difference equation (1.14) admits bifurcations at $\lambda_1^* = \frac{1}{2}$ and $\lambda_2^* = 1$, given by a change in stability of the system, from uniform asymptotic stability for $\lambda < \frac{1}{2}$ to instability for $\lambda > 1$ in two steps. In addition, at both bifurcation points, the number of bounded entire solutions changes significantly.

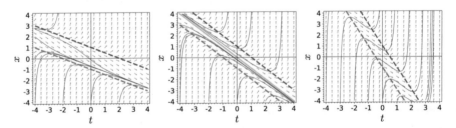

Figure 1.6 Rate-induced tipping from Example 1.2.5 with the quasi-static equilibria $x = -\lambda t - 1$ (attractive, green dashed) and $x = -\lambda t + 1$ (repulsive, red dashed) for $\lambda = \frac{1}{2}$ (left), $\lambda = 1$ (centre, with the semi-stable solution $\varphi_1(t, 0, 0) = -t$ in orange) and $\lambda = 2$ (right)

The following example of rate-induced tipping was discussed in [13, Section 3].

Example 1.2.5 (Rate-Induced Tipping) *Consider the nonautonomous differential equation*

$$\dot{x} = (x + \lambda t)^2 - 1, \tag{1.15}$$

which induces the process φ_λ, depending on a rate $\lambda > 0$, which will be understood as bifurcation parameter. We consider solutions starting in the negative half line at initial time $\tau = 0$, so solutions of the form $\varphi_\lambda(\cdot, 0, \xi)$ with a fixed $\xi \leq 0$. The zeros of the right hand side of (1.15) are given by $x = -\lambda t \pm 1$. These (time-dependent) zeros, the so-called quasi-static equilibria, *are not solutions of the nonautonomous differential equation, but solutions close to a quasi-static equilibrium will track it if they are attractive (i.e. the right hand side has negative derivative). The quasi-static equilibrium $x = -\lambda t - 1$ is attractive, while $x = -\lambda t + 1$ is repulsive. One can show that if $\lambda < 1$, then $\varphi_\lambda(t, 0, \xi)$ tracks the attractive quasi-static equilibrium for all times $t > 0$, meaning that the solution converges to $-\infty$ (see Fig. 1.6 (left)). A rate-induced bifurcation occurs when λ is increased beyond 1. Then the solution $\varphi_\lambda(t, 0, \xi)$ is not able to track the attractive quasi-static equilibrium, since at some point, it will be bigger than the repulsive quasi-static equilibrium, and $\varphi_\lambda(t, 0, \xi)$ converges to ∞ in the limit $t \to \infty$ (see Fig. 1.6 (right)).*

After working with processes so far, we now turn to skew product flows:

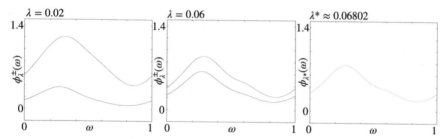

Figure 1.7 Fold bifurcation from Example 1.2.6: Repelling (red) and attracting (green) invariant graphs $\phi_\lambda^\pm$ for the discrete skew product flow with parameters $\alpha = 5, \lambda = 0.02, \lambda = 0.06$ and $\lambda = 0.06802 \approx \lambda^*$, respectively. In this case, there is a single continuous neutral invariant graph ϕ_{λ^*} (orange) at the bifurcation point $\lambda = \lambda^*$

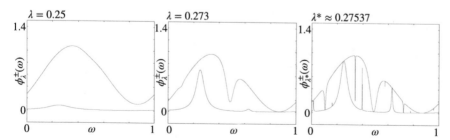

Figure 1.8 Fold bifurcation from Example 1.2.6: Repelling (red) and attracting (green) invariant graphs $\phi_\lambda^\pm$ for the discrete skew product flow with parameters $\alpha = 100, \lambda = 0.25, \lambda = 0.273$ and $\lambda = 0.27537 \approx \lambda^*$, respectively. In this case, a pair of non-continuous pinched invariant graphs exist at the bifurcation point $\lambda = \lambda^*$, often referred to as strange non-chaotic attractors, respectively, repellers

Example 1.2.6 (Fold Bifurcation for Skew Product Flows) *Let $\alpha > 1$ and λ be a bifurcation parameter in $[0, 1]$. Consider a nonautonomous difference equation*

$$x_{t+1} = \arctan(\alpha x_t) - \tfrac{1}{2}(\sin(2\pi\theta(t, \omega)) + 1) - 2\lambda$$

driven by an irrational rotation of the circle, that is,

$$\theta : \mathbb{Z} \times \Omega \to \Omega, \qquad\qquad \theta(t, \omega) := \omega + \varrho t \pmod 1, \qquad (1.16)$$

where $\Omega = \mathbb{R}/\mathbb{Z}$ is the unit circle and $\varrho = \frac{1+\sqrt{5}}{2} \notin \mathbb{Q}$ the golden mean (cf. Example 1.1.1). This problem can be formulated as skew product flow (1.8) on $\Omega \times \mathbb{R}$ with the mapping $\varphi_\lambda(1, \omega, \xi) = \arctan(\alpha\xi) - \tfrac{1}{2}(\sin(2\pi\omega) + 1) - 2\lambda$ and we restrict to parameters $\alpha \in \{5, 100\}$ from now on. The skew product $(\theta, \varphi_\lambda)$ satisfies the criteria (given in [5]) for the occurrence of a nonautonomous subcritical fold bifurcation (of invariant graphs). Thus, there exists a critical parameter $\lambda^* = \lambda^*(\alpha) \in (0, 1)$ such that, depending on the real parameter λ, we obtain the following behaviour:

- *Case $\lambda < \lambda^*(\alpha)$. There exist two invariant graphs $\phi_\lambda^-, \phi_\lambda^+ : \Omega \to \mathbb{R}$ in the region $\Gamma = [0, 2]$. In addition, ϕ_λ^- is repelling, and ϕ_λ^+ is attracting. Note that there also exists a third invariant graph, below the region Γ which is attracting (and persists throughout the parameter range).*
- *Case $\lambda = \lambda^*(\alpha)$. Depending on α, the systems can behave in two different ways:*

  - *If $\alpha = 5$, there exists a semi-stable continuous invariant graph in the region Γ (corresponding to a semi-stable fixed point in the autonomous case of the fold bifurcation); see Fig. 1.7.*
  - *If $\alpha = 100$, there exist two semi-continuous invariant graphs $\phi_\lambda^-, \phi_\lambda^+ : \Omega \to \mathbb{R}$ in Γ with $\phi_\lambda^- \le \phi_\lambda^+$, that are pinched, i.e. there exists $\omega \in \Omega$ such that $\phi_\lambda^-(\omega) = \phi_\lambda^+(\omega)$. In particular, due to the minimality of θ, $\phi_\lambda^-(\omega) = \phi_\lambda^+(\omega)$ on a residual[1] set $R \subseteq \Omega$; see Fig. 1.8. Both R and $\Omega \setminus R$ are invariant and dense, which indicates the complexity of the dynamics.*

- *Case $\lambda > \lambda^*$. There exist no invariant graphs in Γ.*

The subsequent bifurcation phenomena require a state space, which is at least two-dimensional.

Example 1.2.7 (Nonautonomous Hopf Bifurcation) *Let $b : \mathbb{R} \to (0, \infty)$ be a continuous function satisfying the bounds (1.13). Consider the nonautonomous differential equation*

$$\begin{cases} \dot{x} = y + x[\lambda - b(t)(x^2 + y^2)], \\ \dot{y} = -x + y[\lambda - b(t)(x^2 + y^2)], \end{cases} \qquad (1.17)$$

which in polar coordinates decouples into the product system

---

[1] This means that the complement $\Omega \setminus R$ is *meagre* in the sense of [218, p. 78].

$$\begin{cases} \dot{r} = \lambda r - b(t) r^3 \,, \\ \dot{\vartheta} = 1 \,. \end{cases}$$

Since the radial component is the Bernoulli equation (1.12) discussed in Example 1.2.3, we obtain the general solution

$$r_\lambda(t, \tau, \rho) = \frac{e^{\lambda(t-\tau)} \rho}{\sqrt{1 + 2\rho^2 \int_\tau^t b(s) e^{2\lambda(s-\tau)} \, ds}}$$

and

$$\rho_\lambda(t) = \lim_{\tau \to -\infty} r_\lambda(t, \tau, \rho) = \frac{1}{\sqrt{2 \int_{-\infty}^t e^{2\lambda(s-t)} b(s) \, ds}}$$

as pullback limit of the radial solution; the dynamics of the radial component is illustrated in Fig. 1.4. For constant functions $b(t) \equiv \beta > 0$ the planar system (1.17) is a prototype for a Hopf bifurcation, where the asymptotic stability of the origin gets transferred to the periodic orbit $S_\lambda := \{(\xi, \eta) \in \mathbb{R}^2 : \sqrt{\xi^2 + \eta^2} = \sqrt{\frac{\lambda}{\beta}}\}$ as λ increases through the critical value 0; note that $\rho_\lambda(t) \equiv \sqrt{\frac{\lambda}{\beta}}$ on $\mathbb{R}$. Having r_λ explicitly available, one can also see through the situation of a time-dependent b: At $\lambda = 0$ the trivial solution $\mathbb{R} \times \{(0,0)\}$ bifurcates into the nonautonomous set

$$\mathcal{S}_\lambda := \{(t, \xi, \eta) \in \mathbb{R} \times \mathbb{R}^2 : \sqrt{\xi^2 + \eta^2} = \rho_\lambda(t)\} \,.$$

Moreover, the global pullback attractor of (1.17) for parameters $\lambda > 0$ is given by the set $\mathcal{A}_\lambda := \{(t, \xi, \eta) \in \mathbb{R} \times \mathbb{R}^2 : \sqrt{\xi^2 + \eta^2} \leq \rho_\lambda(t)\}$. Since the global pullback attractor for $\lambda \leq 0$ is $\mathcal{A}_\lambda = \mathbb{R} \times \{(0,0)\}$, this nonautonomous Hopf bifurcation can be understood as attractor bifurcation as well.

Already in Example 1.2.4 we have seen that the loss of stability of the trivial solution occurred in two steps. Such a two-step scenario was proposed by L. Arnold [7] and also manifests in the following example, which requires the following notion: A nonempty, compact, and invariant nonautonomous set $\mathcal{A}$ is called a *global pullback attractor* of a skew product flow (θ, ϕ), if

$$\lim_{t \to \infty} d(\varphi(t, \theta(-t, \omega), B), \mathcal{A}(\omega)) = 0 \quad \text{for all } \omega \in \Omega$$

holds for every nonempty bounded subset $B \subset X$.

Example 1.2.8 (Sacker–Neimark Bifurcation for Skew Product Flows) *Assume that $\lambda \geq 0$ and let $\Omega = \mathbb{R}/\mathbb{Z}$. Consider a nonautonomous, planar difference equation*

$$\begin{pmatrix} x_{t+1} \\ y_{t+1} \end{pmatrix} = \frac{\arctan\left(\lambda \sqrt{x_t^2 + y_t^2}\right)}{3\sqrt{2}\sqrt{x_t^2 + y_t^2}} A(\theta(t, \omega)) \begin{pmatrix} x_t \\ y_t \end{pmatrix} , \qquad (1.18)$$

which is trivially extended from $\mathbb{R} \setminus \{(0,0)\}$ to $\mathbb{R}^2$, again driven by the irrational rotation (1.16) and

$$A(\omega) := \begin{pmatrix} 1/\sqrt{2} & 0 \\ 0 & \sqrt{2} \end{pmatrix} \begin{pmatrix} \cos(2\pi\omega) & \sin(2\pi\omega) \\ -\sin(2\pi\omega) & \cos(2\pi\omega) \end{pmatrix}.$$

This gives rise to a skew product flow (1.8) *where* $f : \Omega \times \mathbb{R}^2 \times \Lambda \to \mathbb{R}^2$,

$$f(\omega, x, y, \lambda) = \begin{cases} \dfrac{\arctan(\lambda\sqrt{x^2+y^2})}{3\sqrt{2}\sqrt{x^2+y^2}} A(\omega)\binom{x}{y} & : x^2 + y^2 \neq 0, \\ 0 & : x^2 + y^2 = 0. \end{cases}$$

In order to give a concise description of the bifurcation pattern in this setting, we concentrate on the behaviour of the global attractor. It is easily shown that $\|f(\omega, \xi, \eta, \lambda)\| \leq 1$, *then* $(\theta, \varphi_\lambda)(\{1\} \times \Omega \times \mathbb{R}^2) \subseteq \Omega \times \bar{B}_1(0)$ *so that the global attractor can be defined as* $\mathcal{A}_\lambda = \bigcap_{t \geq 0}(\theta, \varphi_\lambda)(\{t\} \times \Omega \times \bar{B}_1(0))$. *The skew product* $(\theta, \varphi_\lambda)$ *satisfies the criteria [6] for a nonautonomous Sacker–Neimark bifurcation. Hence,* (1.18) *undergoes a full two-step bifurcation scenario, in the sense that there exist two critical* λ-*values, determined by the maximal Lyapunov exponent of* A *given by* $\chi(A) = \frac{\sqrt{2}+1/\sqrt{2}}{2} = \frac{3}{2\sqrt{2}}$ *(see [22, 100]). Then, the critical parameters are given by* $\lambda_1^* = \frac{3\sqrt{2}}{\chi(A)} = 4$ *and* $\lambda_2^* = 3\sqrt{2}\chi(A) = \frac{9}{2}$. *Thus, depending on* $\lambda \geq 0$, *we obtain the following behaviour (Fig. 1.9):*

- *Case* $\lambda < \lambda_1^*$. *The global attractor is* $\Omega \times \{(0,0)\}$.
- *Case* $\lambda_1^* < \lambda < \lambda_2^*$. *There exists at least one* $\omega \in \Omega$ *such that* $\mathcal{A}_\lambda(\omega)$ *is a line segment of positive length.*
- *Case* $\lambda > \lambda_2^*$. *For all* $\omega \in \Omega$ *the fibres* $\mathcal{A}_\lambda(\omega)$ *are closed topological disks and depend continuously on* ω. *Further, the compact* $(\theta, \varphi_\lambda)$-*invariant set* $\mathcal{S}_\lambda = \partial\mathcal{A}_\lambda$ *is the global attractor outside* $\Omega \times \{(0,0)\}$ *in the sense that*

$$\mathcal{S}_\lambda = \bigcap_{t \geq 0}(\theta, \varphi_\lambda)\left(\{t\} \times \Omega \times \left(\bar{B}_1(0) \setminus \bar{B}_\delta(0)\right)\right)$$

for all sufficiently small $\delta > 0$.

The final example also requires the state space to be at least two-dimensional but again concatenates two hyperbolic autonomous problems.

Example 1.2.9 (Transcritical Solution Bifurcation) *Consider the nonlinear difference equation*

$$\begin{pmatrix} x_{t+1} \\ y_{t+1} \end{pmatrix} = \begin{pmatrix} b_t & 0 \\ \lambda & c_t \end{pmatrix} \begin{pmatrix} x_t \\ y_t \end{pmatrix} + \begin{pmatrix} 0 \\ x_t^2 \end{pmatrix} \tag{1.19}$$

with the asymptotically autonomous (indeed piecewise constant) sequences

$$b_t := \begin{cases} 2 & : t < 0, \\ \frac{1}{2} & : t \geq 0, \end{cases} \qquad c_t := \begin{cases} \frac{1}{2} & : t < 0, \\ 2 & : t \geq 0. \end{cases}$$

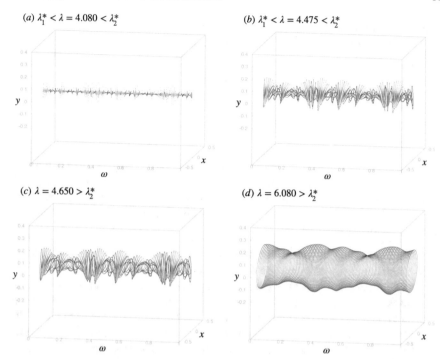

Figure 1.9 Sacker–Neimark bifurcation from Example 1.2.8: The global attractor $\mathcal{A}_\lambda$ of the skew product flow $(\theta, \varphi_\lambda)$. Figures (**a**), (**b**), show $\mathcal{A}_\lambda$ shortly after the first bifurcation parameter λ_1^* and just before the second λ_2^*, respectively. Figure (**c**) shows $\mathcal{A}_\lambda$ shortly after λ_2^* where an invariant torus has just formed, and finally, (**d**) shows the split off invariant torus far from the second bifurcation

The first component of the general solution $\varphi_\lambda(\cdot, 0, \xi, \eta)$ is $\varphi_\lambda^1(t, 0, \xi, \eta) = 2^{-|t|}\xi$. Using the Variation of Constants formula, the second component fulfils

$$\varphi_\lambda^2(t, 0, \xi, \eta) = \begin{cases} 2^t \left(\eta + \frac{4}{7}\xi^2 + \frac{2\lambda}{3}\xi\right) + o(1): & t \to \infty, \\ 2^{-t} \left(\eta - \frac{2}{7}\xi^2 - \frac{2\lambda}{3}\xi\right) + o(1): & t \to -\infty \end{cases}$$

and in conclusion, $\varphi_\lambda(\cdot, 0, \eta)$ is bounded if and only if $\xi = \eta = 0$ or $\xi = -\frac{14}{9}\lambda$, $\eta = \frac{28}{81}\lambda^2$. Therefore, besides the zero solution there is a unique entire bounded solution to (1.19) passing through the initial point (ξ, η) at time $t = 0$ for $\lambda \neq 0$. This means the solution bifurcation pattern sketched in Fig. 1.10 holds.

Figure 1.10 Transcritical
solution bifurcation from Example 1.2.9 with $\lambda^* = 0$
for different parameters λ:
Initial values (ξ, η) guaranteeing a homoclinic solution
$\varphi_\lambda(\cdot, 0, \xi, \eta)$ of equation
(1.19)

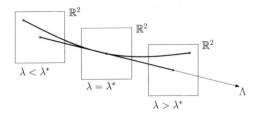

To conclude this section, we observed in our Examples 1.2.2–1.2.9 that parameter variation led to a change in the number of bounded entire solutions for the respective nonautonomous equations—we denote this behaviour as *solution bifurcation*. In the Examples 1.2.2–1.2.4, 1.2.7, 1.2.8 we additionally observed a 'topological' change in the attractor A_λ as follows:

- From a nonautonomous set A_λ consisting of singleton fibres, over the empty set to a repeller as λ was increasing through 1 in Example 1.2.2.
- A continuous transition of A_λ from having singleton to interval fibres as λ increases through the value 1 in Example 1.2.3. The same can be said about the (topological) disks $A_\lambda(t)$ or $A_\lambda(\omega)$ arising in Example 1.2.7, respectively 1.2.8.
- In Example 1.2.4 the trivial solution changes from being an attractor for parameters $\lambda \in (0, \frac{1}{2})$ to a repeller for $\lambda > 1$, while there is no attractor for $\lambda \in (\frac{1}{2}, 1)$.

One can understand such a phenomenon as *attractor bifurcation*; in our examples this also went hand in hand with a change in stability. On the other side, entire solutions can bifurcate while staying unstable.

1.3 Topological Equivalence

For autonomous equations (1.1) a bifurcation is understood as a topological change in their phase portrait when a parameter is varied; that is, there exists no homeomorphism—one speaks of a *topological equivalence*—conjugating phase portraits for different parameter values [148, p. 57, Definition 2.11].

The nonautonomous situation (1.4) is more complicated: On the one hand, a topological equivalence between the solution portraits should be time-dependent, i.e. working with a single homeomorphism acting simultaneously on all fibres of the extended state space is obviously too restrictive. It rather appears more appropriate to work with a family of homeomorphisms $h_t : X \to X$, $t \in \mathbb{T}$, such that at least the solution property is preserved. Thereto, consider two nonautonomous equations

$$\dot{x} = f_i(t, x) \quad \text{or} \quad x_{t+1} = f_i(t, x_t) \quad \text{for } i = 1, 2 \tag{1.20}$$

giving rise to invertible processes $\varphi_i : \mathbb{T} \times \mathbb{T} \times X \to X$. Keeping some instant $\tau \in \mathbb{T}$ fixed, the mapping

$$h_t(x) := \varphi_2(t, \tau, \varphi_1(\tau, t, x))$$

is a homeomorphism with the inverse $h_t^{-1}(x) = \varphi_1(t, \tau, \varphi_2(\tau, t, x))$ satisfying

$$h_t \circ \varphi_1(t, \tau, \cdot) = \varphi_2(t, \tau, \cdot) \circ h_\tau \quad \text{for all } t, \tau \in \mathbb{T}.$$

Thus, this concept of conjugation is too wide, because any two equations (1.20) are equivalent via a family of homeomorphisms. A reasonable nonautonomous notion of topological conjugacy has to fulfil further properties:

- It should be local near a given reference solution.
- Hyperbolicity or nonhyperbolicity properties of the reference solution should be preserved in the sense that the dimension of the stable and unstable manifolds persist, respectively, the dimension of the centre manifold.

On the one hand, [216, 217] or [186, pp. 317ff, Chapter 5] work with the subsequent concept preserving stability properties of the trivial solution: Under the condition $f_i(t, 0) \equiv 0$ for $i = 1, 2$, nonautonomous equations (1.20) are called *topologically equivalent*, provided h_t, h_t^{-1} exist as continuous functions in neighbourhoods of 0 (having open domains being uniform in $t \in \mathbb{T}$) and the limit relations

$$\lim_{x \to 0} h_t(x) = 0, \qquad \lim_{x \to 0} h_t^{-1}(x) = 0 \quad \text{uniformly in } t \in \mathbb{T}$$

hold. On the other hand, [44, 45, 118, 177] require that the norm-distance between h_t and id is globally bounded.

For approaches to topological conjugacy and structural stability in the field of nonautonomous differential equations we refer to [144, 155–157].

Finally, if the mapping h_t is a C^m-diffeomorphism, then one speaks of C^m-*equivalence* [44, 45]. *Local C^m-equivalences* are only defined in (uniform) neighbourhoods of the origin (or another reference solution [216, 217]). For nonautonomous Sternberg theorems yielding C^∞-equivalences we refer to [58].

1.4 Neglected Topics

In this review, we focus on bifurcations of *deterministic* nonautonomous dynamical systems concerning the *asymptotic* (i.e. infinite-time) behaviour, or the structure of a set of specific solutions, and we neither cover bifurcations of random dynamical systems nor finite-time bifurcations. In this section, we give a brief overview about the literature on these neglected topics.

Random dynamical systems are specific nonautonomous dynamical systems that can be expressed as a skew product flow (θ, φ), as in Sect. 1.1.2, but with the difference that $\theta : \mathbb{T} \times \Omega \to \Omega$ is a measurable flow on a probability space $(\Omega, \mathfrak{A}, \mu)$, where μ is an invariant measure with respect to θ, see the monograph [7]. Despite the similar setting, and the fact that many objects used in nonautonomous dynamical systems (such as pullback attractors) have a natural counterpart for random dynamical systems, many results on bifurcations for random dynamical systems do not have an immediate translation to nonautonomous dynamical systems, and vice versa.

Research on bifurcations in random dynamical system started to develop in the 1980s and the main activity first concentrated on statistical quantities using the Markov semigroup. It was explored how stationary measures change under variation of parameters, see the detailed discussions in [7, Chapter 9], [105] and [220]. More recent approaches considered Markov densities [65] and moment maps [23].

Starting from the 1990s, more results on bifurcations took the full skew product flow structure into account and thus described dynamical rather than statistical quantities that undergo changes. First results analysed so-called D-bifurcations which concern bifurcations of invariant measures of the random dynamical system [10, 56, 223]. In [55], it was shown that D-bifurcations do not occur in one-dimensional random systems with additive noise, which follows from synchronisation properties in this setting, see [80, 81] for extensions to higher-dimensional systems. Hopf bifurcations of random dynamical systems were first studied in [9, 26, 211], and recently, Hopf bifurcations were studied that are induced by changes in the sign of the top Lyapunov exponent in the presence of shear [66, 75, 159, 225, 227]. Bifurcations involving a localised version of the top Lyapunov exponent have been analysed in [76], and in [40, 67], bifurcations are analysed that involve a breakdown of a random version of topological equivalence. Bifurcations involving changes in the dichotomy spectrum have been analysed in [40, 69].

In the last fifteen years, bifurcations of random dynamical systems involving bounded noise have been studied extensively. Bounded noise is often a more realistic modelling assumption, since physical quantities only vary within certain bounds. The modelling via bounded noise has profound consequences for the dynamics, since random dynamical systems with unbounded noise have often only one attractor that is supported on the whole state space, and the use of bounded noise makes the distinction of qualitative changes possible that manifest themselves through interactions between localised objects. First studies focussed on one- and two-dimensional scenarios in both discrete and continuous time [37, 104, 230, 231], and relationships to bifurcations in set-valued dynamical systems have been exploited subsequently [145, 149].

Finally, it should be noted that nonautonomous dynamical systems are extremely important for applications in fluid dynamics and oceanography, where a nonautonomous differential equation describes the time-dependent velocity field around an airfoil or of a stretch of ocean surface. For the analysis of such systems, the asymptotic theory as described in this work is not meaningful, and instead, dynamical quantities that describe finite-time behaviour need to be considered. For instance, in order to analyse patterns which influence transport phenomena, Lagrangian coherent structures are important, which can be detected as ridges of the field of finite-time Lyapunov exponents [97] and using spectral methods that yield almost invariant sets [85]. It should be noted that a theory of finite-time bifurcations in such applications has just started to develop, and one very topical application is the understanding of the Antarctic polar vortex in 2002 [35, 166].

Remarks

History Nonautonomous bifurcations were first studied in the context of quasi-periodic differential equations in [47, 214]. Bifurcations of almost periodic solutions to such differential equations can be traced back to the monograph [142] and [134], while corresponding results in discrete time appeared later in [98].

As another early contribution, [140] presents a phenomenological approach to bifurcations in nonautonomous ordinary differential equations. The references [150–152] understand bifurcations as changes in (pullback) stability notions for scalar ordinary differential equations. In contrast, [198–200] describes bifurcations in terms of attraction/repulsion properties of solutions to scalar differential and difference equations. Topological methods were employed in [51] to address bifurcations in control systems. Using a skew product language, the contribution [173] gives elegant nonautonomous counterparts to the classical bifurcation patterns for scalar differential equations.

The references [120, 124] investigate Hopf bifurcations along nonperiodic solutions and [77, 78, 122] apply averaging techniques to deduce nonautonomous counterparts of transcritical and saddle-node (fold) bifurcations. A two-step bifurcation scenario significantly different from [189] was investigated in [121].

Finally, these notes extend our previous surveys [137, Chapter 7], [139, Chapter 8] or [188] and provide a broader, up-to-date scope and various additional examples and results.

Pullback Convergence A prominent form of attraction considered here is denoted as *pullback attraction* and dates back to at least [133]. This concept is central in nonautonomous dynamics and already the Examples 1.2.1, 1.2.2, 1.2.3, and 1.2.7 illustrated that pullback limits

$$\lim_{\tau \to -\infty} \varphi(t, \tau, \xi)$$

yield dynamically more meaningful concepts than the forward limits $t \to \infty$. Such convergence guarantees, for instance, that limit sets become invariant and inherit various canonical properties from their autonomous special cases (cf. [36], [43, pp. 3ff, Chapter 1], [139, pp. 1ff, Chapters 1–3], [141] or [186, pp. 1ff, Chapter 1]). Yet, we do not conceal the fact that pullback convergence strongly emphasises past behaviour and lacks to capture forward dynamics (see the note [138] for a more detailed discussion). A comparison of different attractor notions can be found in [46]. Our repeller concept is taken from [199, p. 13, Definition 2.6]. As a general source for nonautonomous dynamical systems, we refer to the recent monographs [43, 139, 141] or the survey [137] with a focus on discrete dynamics.

Examples and Applications The examples in this text are didactically motivated and thus not intended to cover real-world applications. A satisfactory treatment of such problems is indeed expected to be rather involved. For instance, it requires numerical methods to compute invariant projectors and to approximate dynamical spectra, as well as continuation techniques for entire solutions or invariant graphs. Nevertheless, we refer the reader to, for example, [110, 136].

Part I
Nonautonomous Differential Equations

In this first part, we deal with continuous time $\mathbb{T} = \mathbb{R}$. Our interest is focussed on nonautonomous differential equations

$$\dot{x} = f(t, x) \tag{D}$$

with a continuous right hand side $f : \mathcal{D} \to \mathbb{R}^d$ defined on an open nonautonomous set $\mathcal{D} \subseteq \mathbb{R} \times \mathbb{R}^d$ and satisfying assumptions guaranteeing existence and uniqueness of solutions [96, 146]. Hence, a *general solution* φ yielding the process properties

$$\varphi(\tau, \tau, \xi) = \xi \quad \text{for all } (\tau, \xi) \in \mathcal{D},$$
$$\varphi(t, \tau, \xi) = \varphi(t, s, \varphi(s, \tau, \xi)) \quad \text{for all } t \geq s \geq \tau \text{ and } \xi \in \mathcal{D}(\tau)$$

is at hand. This is guaranteed, e.g. if f fulfils a *local Lipschitz condition* of the form such that for every compact $K \subset \mathcal{D}$ there exists a *Lipschitz constant* $L_K \geq 0$ with

$$\|f(t, x_1) - f(t, x_2)\| \leq L_K \|x_1 - x_2\| \quad \text{for all } (t, x_1), (t, x_2) \in K.$$

Moreover, the general solution $\varphi(\cdot, \tau, \xi)$ exists on $[\tau, \infty)$ under global Lipschitz, linear boundedness or dissipativity conditions (e.g. [96, p. 32, Corollary 6.3]). We are interested in the dynamics of (D) on an unbounded interval $\mathbb{I} \subseteq \mathbb{R}$, which is typically one of the axes $\mathbb{R}_0^+$, $\mathbb{R}_0^-$ or $\mathbb{R}$.

A special case of (D) on $\mathcal{D} = \mathbb{I} \times \mathbb{R}^d$ is linear, nonautonomous differential equations. Precisely, suppose $A : \mathbb{I} \to \mathbb{R}^{d \times d}$ is continuous, and consider

$$\dot{x} = A(t)x \tag{L}$$

with the *transition matrix* $\Phi(t, s) \in GL(\mathbb{R}^d)$, which satisfies

$$\Phi(\tau, \tau) = \text{id}, \qquad \Phi(t, s)\Phi(s, \tau) = \Phi(t, \tau) \quad \text{for all } \tau, s, t \in \mathbb{I}.$$

Finally, bifurcation theory deals with parametrised equations

$$\dot{x} = f(t, x, \lambda), \tag{D_λ}$$

depending on a parameter λ from some ambient set Λ. Here, $f : \mathcal{D} \times \Lambda \to \mathbb{R}^d$ may fulfil the above assumptions for every $\lambda \in \Lambda$. The general solution of (D_λ) is denoted by φ_λ. It inherits its smoothness (in τ, ξ and λ) from the right hand side of (D_λ) (see [48, p. 89, Theorem 1.1]).

Chapter 2
Spectral Theory, Stability and Continuation

Bifurcations for autonomous dynamical systems cannot be explained using linear theory alone, but eigenvalues or Floquet multipliers of linearisations are important for the analysis of stability and the formulation of necessary conditions for bifurcations, while generalised eigenspaces shape the skeleton of the dynamics [53, 102]. The importance of linear systems extends to nonautonomous dynamical systems, but here eigenvalues lose their meaning (as demonstrated for the periodic differential equation (2.1) below), and we need different spectral concepts in order to arrive at an appropriate notion of hyperbolicity or even to determine exponential stability. In this chapter, we introduce the notions of exponential dichotomy [139] and the related dichotomy spectrum, as well as Lyapunov exponents [24] as appropriate generalisations of eigenvalues. These central concepts of the linear theory are compared, and important properties are provided. Using the dichotomy spectrum, we also present continuation results that rule out local nonautonomous bifurcations in a vicinity of a reference solution, and we describe a constructive method to approximate continuations of this reference solution at least locally.

2.1 Spectral Theory

Linear systems canonically occur as variational equations along given reference solutions of ordinary differential equations and are typically nonautonomous if the reference solution is nonconstant (even if the differential equation is autonomous).

Differing from linear autonomous systems $\dot{x} = Ax$, where the eigenvalues of the matrix A determine the stability, the time-dependent eigenvalues of the matrices $A(t)$ in (L) cannot be used for stability investigations. We demonstrate this by means of the π-periodic linear system

$$\dot{x} = \begin{pmatrix} -1 + \frac{3}{2}\cos^2 t & 1 - \frac{3}{2}\cos t \sin t \\ -1 - \frac{3}{2}\cos t \sin t & -1 + \frac{3}{2}\sin^2 t \end{pmatrix} x \tag{2.1}$$

© The Author(s), under exclusive license to Springer Nature Switzerland AG 2023
V. Anagnostopoulou et al., *Nonautonomous Bifurcation Theory*, Frontiers
in Applied Dynamical Systems: Reviews and Tutorials 10,
https://doi.org/10.1007/978-3-031-29842-4_2

from [96, p. 121, Example 7.1]. The coefficient matrix of (2.1) has the constant eigenvalues $\frac{-1\pm\sqrt{7}i}{4}$ with negative real parts. However, this does not allow us to conclude asymptotic stability, since $\phi(t) := e^{t/2}\left(\begin{smallmatrix} -\cos t \\ \sin t \end{smallmatrix}\right)$ is an unbounded solution.

Despite this example, the periodic case is well understood using Floquet theory, which says that for a T-periodic linear system (L), the stability is determined by the *Floquet multipliers*, i.e. the eigenvalues of the *monodromy matrix* $\Pi := \Phi(T,0)$. In particular, with $\sigma(\Pi) \subseteq \mathbb{C}$ denoting the set of eigenvalues of Π, the periodic linear system (L) is uniformly asymptotically stable if $\sigma(\Pi) \subset B_1(0)$, and (L) is unstable if $\sigma(\Pi) \not\subset \bar{B}_1(0)$, see [96, p. 120, Theorem 7.2]. Note that the Floquet multipliers of (2.1) are given by $e^{-2\pi}$ and e^{π} and thus indicate instability.

In this chapter, we aim to study time-dependent linear systems (L) beyond periodicity, and this leads to the following question:

> *Given that the eigenvalues cannot be used to analyse*
> *the stability of nonautonomous linear systems,*
> *what alternative tools are available?*

A historically first approach to this question is the concept of *characteristic* or *Lyapunov exponents* yielding criteria for asymptotic stability. Yet, as classical examples show, without a regularity assumption, Lyapunov exponents do not provide a robust stability notion in the sense that asymptotic stability of a linear equation can be destroyed by perturbations of order $o(x)$ in (L) [183, § 2]. Beyond questions related to stability, we also aim at a nonautonomous version of hyperbolicity.

In this section, we discuss two approaches to study spectral properties of nonautonomous differential equations: the *dichotomy spectrum* (due to R.J. Sacker and G.R. Sell [210, 215]) and the *Lyapunov spectrum* (due to A.M. Lyapunov [1]). We find the dichotomy spectrum more suitable when dealing with processes, since for general time-dependencies, it is difficult to check regularity (see Appendix A.1). When working with skew product flows, the particular driving dynamics often allows to infer regularity, and for this reason, the approach via Lyapunov exponents and the Multiplicative Ergodic Theorem is advantageous in this context.

2.1.1 Dichotomy Spectrum

In this subsection, we introduce the concept of a dichotomy spectrum. The definition of the dichotomy spectrum is based on the crucial notion of an exponential dichotomy. Exponential dichotomies [54, 59] form a concept of hyperbolicity for nonautonomous dynamical systems and are fundamental for the definition of the dichotomy spectrum [201, 210, 215].

Let $\mathbb{I} \subseteq \mathbb{R}$ be an unbounded interval. An *invariant projector* for the linear system (L) is a function $P : \mathbb{I} \to \mathbb{R}^{d\times d}$ of projections $P(t) = P(t)^2$ such that $\Phi(t,\tau)P(\tau) = P(t)\Phi(t,\tau)$ for all $t, \tau \in \mathbb{I}$. It follows that all ranges $R(P(t))$ have the same dimension for all $t \in \mathbb{I}$, and the same holds for all null spaces $N(P(t))$.

Definition 2.1.1 (Exponential Dichotomy) *The linear system (L) is said to admit an* exponential dichotomy *on $\mathbb{I}$ if there exist an invariant projector P and real numbers $K \geq 1$ and $\beta > 0$ such that*

$$\|\Phi(t,s)P(s)\| \leq Ke^{\beta(s-t)}, \quad \|\Phi(s,t)(\mathrm{id} - P(t))\| \leq Ke^{\beta(s-t)} \quad \textit{for all } s \leq t \,.$$

Behind this formal definition stands the following geometric idea concerning the *stable nonautonomous set*:

$$\mathcal{V}^+ := \left\{ (\tau, \xi) \in \mathbb{I} \times \mathbb{R}^d : \lim_{t \to \infty} \Phi(t, \tau)\xi = 0 \right\}$$

(defined if $\mathbb{I}$ is unbounded above) and the *unstable nonautonomous set*

$$\mathcal{V}^- := \left\{ (\tau, \xi) \in \mathbb{I} \times \mathbb{R}^d : \lim_{t \to -\infty} \Phi(t, \tau)\xi = 0 \right\}$$

(defined if $\mathbb{I}$ is unbounded below), see also Fig. 2.1. The *Whitney sum* of these two sets forms a hyperbolic splitting $\mathbb{I} \times \mathbb{R} = \mathcal{V}^+ \oplus \mathcal{V}^-$ of the extended state space. Summarising results from [54, p. 19] and [59, pp. 162ff, §3] leads to the next result.

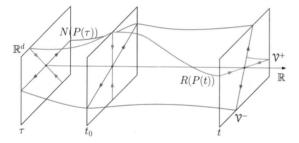

Figure 2.1 Fibres for the stable set $\mathcal{V}^+$ (green) and the unstable set $\mathcal{V}^-$ (red) of a linear system (L) with an exponential dichotomy, intersecting along the trivial solution

Proposition 2.1.2 (Characterisation of $\mathcal{V}^+$ and $\mathcal{V}^-$) *Assume that a linear system (L) admits an exponential dichotomy on $\mathbb{I}$ with an invariant projector P. If $\mathbb{I}$ is unbounded above, then the stable nonautonomous set $\mathcal{V}^+$ satisfies $R(P(t)) = \mathcal{V}^+(t)$ for all $t \in \mathbb{I}$. Similarly, for $\mathbb{I}$ unbounded below, the unstable nonautonomous set $\mathcal{V}^-$ fulfils $N(P(t)) = \mathcal{V}^-(t)$ for all $t \in \mathbb{I}$. Thus, for exponential dichotomies on $\mathbb{R}$, the invariant projector P is uniquely determined.*

In contrast, it can be shown that the null space of invariant projectors for an exponential dichotomy on $\mathbb{I} = \mathbb{R}_0^+$ and the range of invariant projectors for an exponential dichotomy on $\mathbb{I} = \mathbb{R}_0^-$ are not uniquely determined [201, Remark 5.6].

Example 2.1.3 (Exponential Dichotomies for Autonomous Linear Systems) *An autonomous linear differential equation*

$$\dot{x} = Ax$$

possesses an exponential dichotomy on an unbounded interval, if and only if it is hyperbolic, i.e. the matrix $A \in \mathbb{R}^{d \times d}$ has no eigenvalues on the imaginary axis. For a spectral splitting $\sigma(A) = \sigma^+ \cup \sigma^-$ with $\operatorname{Re} \sigma^+ \subset (-\infty, 0)$ and $\operatorname{Re} \sigma^- \subset (0, \infty)$, the nonautonomous sets $\mathcal{V}^+$ and $\mathcal{V}^-$ have constant fibres and are given by the direct sum of the generalised eigenspaces corresponding to eigenvalues in σ^+ and σ^-, respectively. A similar result holds for periodic linear differential equations (L) using a spectral splitting for the monodromy matrix Π.

It is well known [102, p. 157, Theorem 3] that hyperbolicity is a generic property in the class of autonomous or periodic linear systems, i.e. hyperbolic systems are open and dense among autonomous and periodic systems in $\mathbb{R}^d$. When passing to nonautonomous linear systems, the situation is more subtle. The Roughness Theorem ensures that an exponential dichotomy of (L) on an unbounded interval $\mathbb{J}$ extends to linearly homogeneous perturbations $\dot{x} = [A(t) + B(t)]x$, provided the function $B : \mathbb{J} \to \mathbb{R}^{d \times d}$ is globally bounded with a small norm for $\mathbb{J} = \mathbb{R}$ [126, Theorem 3.2] resp. has a small norm on an unbounded subinterval of $\mathbb{J}$ for a half line $\mathbb{J}$ [126, Theorem 3.1]. As a result, on the entire line $\mathbb{J} = \mathbb{R}$, an exponential dichotomy is merely an open property. However, systems with an exponential dichotomy are not dense and consequently not generic in linear systems with bounded coefficient matrices [178].

The following result from [54] links an exponential dichotomy on $\mathbb{R}$ to the exponential dichotomies on both half lines.

Theorem 2.1.4 (Characterisation of Exponential Dichotomies) *The linear system* (L) *admits an exponential dichotomy on $\mathbb{R}$ if and only if the following two conditions are fulfilled:*

(a) (L) *admits exponential dichotomies on both $\mathbb{R}_0^+$ and $\mathbb{R}_0^-$ with respective projectors P^+ and P^-*

(b) $R(P^+(0)) \oplus N(P^-(0)) = \mathbb{R}^d$.

We next introduce an appropriate spectral concept that is based on the notion of an exponential dichotomy. For $\gamma \in \mathbb{R}$, consider the shifted equation

$$\dot{x} = \big(A(t) - \gamma \operatorname{id}\big)x, \qquad\qquad (L_\gamma)$$

and we will be interested for which γ, the shifted system (L_γ) admits an exponential dichotomy. We also say that (L_∞) admits an exponential dichotomy if there exists a $\gamma \in \mathbb{R}$ such that (L_γ) admits an exponential dichotomy with projector $P_\gamma(t) \equiv \operatorname{id}$. Accordingly, we say that $(L_{-\infty})$ admits an exponential dichotomy, if there exists a $\gamma \in \mathbb{R}$ such that (L_∞) admits an exponential dichotomy with growth rate γ and projector $P_\gamma(t) \equiv 0$.

Definition 2.1.5 (Dichotomy Spectrum) *The dichotomy spectrum of (L) on the unbounded interval $\mathbb{I}$ is defined by*

$$\Sigma_{\mathbb{I}}(A) := \big\{\gamma \in \mathbb{R} \cup \{\pm\infty\} : (L_\gamma) \text{ does not have an exponential dichotomy on } \mathbb{I}\big\}.$$

Special cases are the sets

$$\Sigma(A) := \Sigma_{\mathbb{R}}(A), \qquad \Sigma^+(A) := \Sigma_{[0,\infty)}(A), \qquad \Sigma^-(A) := \Sigma_{(-\infty,0]}(A);$$

the latter two are called the forward *and* backward dichotomy spectrum *of* (L), *respectively, but we often refer to them simply as dichotomy spectra.*

The forward and backward dichotomy spectra are subsets of $\Sigma(A)$. Moreover, dichotomy spectra are unions of finitely many intervals [201, 215].

Theorem 2.1.6 (Spectral Theorem) *The dichotomy spectrum $\Sigma_{\mathbb{I}}(A)$ of (L) on the unbounded interval $\mathbb{I}$ is of the form*

$$\Sigma_{\mathbb{I}}(A) = [a_1, b_1] \cup \cdots \cup [a_n, b_n],$$

where $-\infty \leq a_1 \leq b_1 < a_2 \leq b_2 < \cdots < a_n \leq b_n \leq \infty$, and $n \in \{1, \ldots, d\}$. Define the invariant projectors $P_{\gamma_0} := 0$, $P_{\gamma_n} := \mathrm{id}$, and for $1 \leq i < n$, choose $\gamma_i \in (b_i, a_{i+1})$ and projectors P_{γ_i} of the exponential dichotomy of (L_{γ_i}) on $\mathbb{I}$. Then the invariant nonautonomous sets $\mathcal{V}_i$, defined by

$$\mathcal{V}_i(t) := R(P_{\gamma_i}(t)) \cap N(P_{\gamma_{i-1}}(t)) \quad \text{for all } i \in \{1, \ldots, n\} \text{ and } t \in \mathbb{I}$$

are the so-called spectral manifolds *such that we have the* Whitney sum

$$\mathcal{V}_1 \oplus \cdots \oplus \mathcal{V}_n = \mathbb{I} \times \mathbb{R}^d,$$

and $\mathcal{V}_i \neq \mathbb{I} \times \{0\}$ for $i \in \{1, \ldots, n\}$.

The spectral intervals $[a_i, b_i]$ can be considered as a nonautonomous counterpart to the eigenvalue real parts for autonomous problems (see below). If all spectral intervals are singletons, one speaks of a *discrete spectrum*. In addition, the nonautonomous sets $\mathcal{V}_i$ are nonautonomous counterparts to the sum of generalised eigenspaces for eigenvalues with the same real parts (see [102, pp. 110ff, Chapter 6] and Example 2.1.3). The fibres of each of these sets are linear subspaces and have constant dimension $\dim \mathcal{V}_i$, called *multiplicity* of the corresponding spectral interval $[a_i, b_i]$ for $i \in \{1, \ldots, n\}$.

As a sufficient condition to exclude super-exponential growth or decay of solutions, we say a linear system (L) has *bounded growth*, if there exist reals $K \geq 1$ and $\alpha \geq 0$ such that

$$\|\Phi(t, s)\| \leq K e^{\alpha |t-s|} \quad \text{for all } t, s \in \mathbb{I}.$$

Note that the inclusion $\Sigma(A) \subseteq [-\alpha, \alpha]$ follows for systems with bounded growth, as well as $\Sigma(A) \neq \emptyset$. It is not difficult to show that systems (L) with bounded coefficient matrix A have bounded growth.

Similarly to the situation in Proposition 2.1.2, in the case $\mathbb{I} = \mathbb{R}$, a dynamical characterisation can be achieved for the spectral manifolds $\mathcal{V}_i$. This establishes a 'nonautonomous linear algebra'. Given a growth rate $\gamma \in \mathbb{R}$, we define the

- γ-*stable nonautonomous set*

$$\mathcal{V}_\gamma^+ := \left\{ (\tau, \xi) \in \mathbb{R} \times \mathbb{R}^d : \sup_{\tau \leq t} \|\Phi(t, \tau)\xi\| e^{\gamma(\tau - t)} < \infty \right\}$$

- and the *γ-unstable nonautonomous set*

$$\mathcal{V}_\gamma^- := \left\{ (\tau, \xi) \in \mathbb{R} \times \mathbb{R}^d : \sup_{t \leq \tau} \|\Phi(t, \tau)\xi\| e^{\gamma(\tau - t)} < \infty \right\},$$

whose fibres are linear subspaces of $\mathbb{R}^d$. In particular, we have $\mathcal{V}^\pm = \mathcal{V}_0^\pm$. Similar to the situation in Theorem 2.1.6, choose rates $\gamma_i \in (b_i, a_{i+1})$ for $i \in \{1, \dots, n-1\}$. It follows that $\mathcal{V}_1 = \mathcal{V}_{\gamma_1}^+$, $\mathcal{V}_n = \mathcal{V}_{\gamma_n}^-$, and

$$\mathcal{V}_i = \mathcal{V}_{\gamma_i}^+ \cap \mathcal{V}_{\gamma_{i-1}}^- \qquad \text{for all } i \in \{2, \dots, n-1\}.$$

Note that such a characterisation is not possible if $\mathbb{I}$ is not the entire real line. If $\mathbb{I}$ is a half line, then only one of the nonautonomous sets $\mathcal{V}_i$ is uniquely determined. This follows from non-uniqueness of the projectors as indicated in Proposition 2.1.2.

We now provide dichotomy spectra of linear systems (L) in several special cases:

- *One-dimensional case.* Suppose that $d = 1$ and $A : \mathbb{I} \to \mathbb{R}$ is bounded. Then $\Sigma_\mathbb{I}(a) = [\underline{\beta}_\mathbb{I}(a), \overline{\beta}_\mathbb{I}(a)]$ (see Appendix A.2 for the definition of the *Bohl exponents* $\underline{\beta}_\mathbb{I}(a)$ and $\overline{\beta}_\mathbb{I}(a)$).
- *Autonomous case.* If the matrix function $A : \mathbb{R} \to \mathbb{R}^{d \times d}$ is constant, then $\Sigma(A) = \Sigma^+(A) = \Sigma^-(A) = \operatorname{Re} \sigma(A)$, which results from Example 2.1.3.
- *T-periodic case.* Suppose the matrix function $A : \mathbb{R} \to \mathbb{R}^{d \times d}$ is T-periodic with the monodromy matrix $\Pi = \Phi(T, 0) \in GL(\mathbb{R}^d)$. Then one also has coinciding spectra $\Sigma(A) = \Sigma^+(A) = \Sigma^-(A) = \frac{1}{T} \ln |\sigma(\Pi)|$.
- *Asymptotically autonomous case.* In case the limit $A^+ := \lim_{t \to \infty} A(t)$ or $A^- := \lim_{t \to -\infty} A(t)$ exists, then the spectra become $\Sigma^+(A) = \operatorname{Re} \sigma(A^+)$ and $\Sigma^-(A) = \operatorname{Re} \sigma(A^-)$, respectively.

2.1.2 Lyapunov Spectrum

In this subsection, we consider the notion of a Lyapunov spectrum for linear systems (L), which is the union of all Lyapunov exponents, either forward in time or backward in time. Lyapunov exponents have been introduced by the Russian mathematician and engineer A.M. Lyapunov in his PhD Thesis [162], see also [1, 153].

Definition 2.1.7 (Lyapunov Exponent) *Consider the linear system (L). Then, given a $\xi \in \mathbb{R}^d \setminus \{0\}$, the forward Lyapunov exponent of the solution $t \mapsto \Phi(t, 0)\xi$ is given by*

$$\chi(\xi) := \limsup_{t \to \infty} \frac{1}{t} \ln \|\Phi(t, 0)\xi\|.$$

We often speak of simply a Lyapunov exponent $\chi(\xi)$ *and denote it as* sharp *if the above* lim sup *can be replaced by* lim.

A solution to (L) starting in $\xi \in \mathbb{R}^d \setminus \{0\}$ with $\chi(\xi) < 0$ decays to 0 exponentially as $t \to \infty$. For scalar equations (L), this definition coincides with the one from Appendix A.2. It can be shown [1, Corollary 2.3.1] that there are only finitely many Lyapunov exponents, which form the so-called Lyapunov spectrum.

Definition 2.1.8 (Lyapunov Spectrum) *Consider the linear system (L). The forward Lyapunov spectrum is given by the finite set*

$$\Sigma_{lyap}(A) := \bigcup_{0 \neq \xi \in \mathbb{R}^d} \{\chi(\xi)\} .$$

Similar to the dichotomy spectrum, the Lyapunov spectrum measures exponential growth behaviour of the linear system (L), but there are important differences. While the Lyapunov spectrum directly measures characteristic growth behaviour of solutions, the dichotomy spectrum is based on the notion of an exponential dichotomy, which describes hyperbolicity in the context of nonautonomous dynamical systems. Having an exponential dichotomy basically means that zero exponential growth is not observed, but this is stronger than not having zero in the Lyapunov spectrum. Hence, the dichotomy spectrum, which is an interval spectrum, is always a superset of the Lyapunov spectrum [199, Theorem 4.30].

Proposition 2.1.9 (Lyapunov and Dichotomy Spectra) *The forward Lyapunov spectrum of (L) is a subset of the forward dichotomy spectrum, i.e. $\Sigma_{lyap}(A) \subseteq \Sigma^+(A)$.*

We note that due to non-uniqueness of the corresponding spectral manifolds from Theorem 2.1.6 and due to the fact that for a specific Lyapunov exponent $\hat{\chi}$, the set $V(\hat{\chi}) := \{x \in \mathbb{R}^d : \chi(x) = \hat{\chi}\}$ is not a linear space in general, and the spectral manifolds cannot be directly linked to spaces associated with Lyapunov exponents. However, it is possible to show that for the smallest Lyapunov exponent $\chi_s \in \Sigma_{lyap}(A)$, the set $V(\chi_s)$ is linear and a subspace of $\mathcal{V}_1(0)$ from Theorem 2.1.6.

We note that Lyapunov exponents and spectra also exist on intervals $\mathbb{R}_\tau^-$, which either follows analogously or via time reversal. This means that an analogous result to Proposition 2.1.9 holds, relating this *backward Lyapunov spectrum* to $\Sigma^-(A)$. Since such results are of minor importance here, from now on we call $\Sigma_{lyap}(A)$ simply the *Lyapunov spectrum*.

As outlined in Sect. 2.2 below, negativity of the dichotomy spectrum directly implies exponential stability under nonlinear perturbations. This is not true in general for Lyapunov exponents, and for this reason, Lyapunov exponents have not played an important role in results on nonautonomous bifurcations for processes so far.

This is different in the setting of skew product flows. Thereto, suppose that a dynamical system $\theta : \mathbb{R} \times \Omega \to \Omega$ on a compact metric space Ω drives a nonautonomous ordinary differential equation

$$\dot{x} = f(\theta(t,\omega), x) \,. \tag{D^ω}$$

Under natural smoothness assumptions on f, it induces a skew product flow (θ, φ). In this context, the existence of strict Lyapunov exponents for the resulting variational equation is highly nontrivial and results from the Multiplicative Ergodic Theorem. We do not discuss this in the present continuous-time setting but refer the reader to analogous results in Sect. 5.1.2 for discrete time.

For our purposes in Sect. 3.3 below, it is sufficient to restrict to scalar differential equations (D^ω). Consider a (θ, φ)-invariant graph $\phi : \Omega \to \mathbb{R}$, and assume that Ω is equipped with an ergodic invariant measure μ and that $\ln^+ |D_2 f(\cdot, \phi(\cdot))|$ is μ-integrable.[1] Then Birkhoff's ergodic theorem [169, pp. 459ff] guarantees the almost sure existence of the *Lyapunov exponent*

$$\chi_\mu(\phi) := \lim_{t\to\infty} \frac{1}{t} \int_0^t D_2 f(\theta(t,\omega), \phi(\omega)) \, \mathrm{d}s \quad \text{for } \mu\text{-almost all } \omega \in \Omega \,,$$

as well as the characterisation

$$\chi_\mu(\phi) = \int_\Omega D_2 f(\omega, \phi(\omega)) \, \mathrm{d}\mu(\omega) \,.$$

2.2 Stability

This section deals with applications of the above dynamical spectra to stability theory for nonlinear equations (D). Given an unbounded interval $\mathbb{I}$, we say that a solution $\phi^* : \mathbb{I} \to \mathbb{R}^d$ of (D) is *hyperbolic* (on $\mathbb{I}$), if the variational equation

$$\dot{x} = D_2 f(t, \phi^*(t)) x \tag{V}$$

has an exponential dichotomy on $\mathbb{I}$. We demonstrate now that the dichotomy spectrum determines stability properties of solutions. We write $\Sigma^+(\phi^*)$ and $\Sigma_{\mathbb{I}}(\phi^*)$ for the dichotomy spectra of (V) and obtain the following stability criteria as illustrated in Fig. 4.2. For a proof, we refer to Theorem A.1.1 in the Appendix.

Proposition 2.2.1 (Dichotomy Spectrum and Linearised Stability) *Consider the differential equation (D) with solution ϕ^* and corresponding variational equation (V). If we assume that $\mathbb{I}$ is bounded below and*

$$\sup_{t\in\mathbb{I}} \|D_2 f(t, \phi^*(t))\| < \infty, \quad \lim_{x\to 0} \sup_{t\in\mathbb{I}} \|D_2 f(t, \phi^*(t) + x) - D_2 f(t, \phi^*(t))\| = 0 \,,$$

then the following statements hold (Fig. 2.2):

---

[1] $\ln^+$ denotes the positive part of the natural logarithm.

Figure 2.2 Form of the dichotomy spectra required in Proposition 2.2.1: (**a**) Uniform exponential stability and (**b**) instability for a spectral interval σ in $(0, \infty)$

(a) $\max \Sigma^+(\phi^*) < 0$ *if and only if* ϕ^* *is uniformly exponentially stable on every half line* $\mathbb{R}_\tau^+$, $\tau \in \mathbb{I}$.

(b). *If a spectral interval* σ *of* $\Sigma^+(\phi^*)$ *fulfils* $\min \sigma > 0$, *then* ϕ^* *is unstable.*

It is clear that this proposition canonically generalises the well-known stability conditions in an autonomous and periodic context [1, p. 84, Theorem 4.2.1]. Yet, the nonhyperbolic situation $0 \in \Sigma_\mathbb{I}(\phi^*)$ is not covered and requires to investigate higher order terms of (D). We refer to the corresponding nonautonomous centre manifold theory as described in Sect. 4.1 and the reduction principle in Theorem 4.1.1.

Sufficient conditions for (not necessarily uniform) exponential stability based on Lyapunov exponents are given as Theorem A.1.3 in the Appendix. We moreover point out that $\max \Sigma^+(\phi^*)$ is an upper bound for the Lyapunov exponents, which can be strictly larger than the top Lyapunov exponent [101, p. 259, Example 3.3.12].

2.3 Continuation

Beyond its role in stability theory (as discussed in Sect. 2.2), this section illustrates that the dichotomy spectrum provides a reasonable hyperbolicity notion for nonautonomous dynamical systems also in the context of bifurcation theory. Here, hyperbolicity excludes bifurcations for a large class of possible perturbations of a nonautonomous dynamical system. It will become clearer below that for such kind of perturbations, it is advantageous to allow the parameter space Λ to be an open subset of a Banach space Y.[2]

Throughout this section, we consider parameter-dependent equations (D_λ) on the entire line $\mathbb{R}$ and assume the following on the smoothness of the right hand side.

Hypothesis 2.3.1 *Let* $m \in \mathbb{N}$. *Suppose that the function* $f : \mathcal{D} \times \Lambda \to \mathbb{R}^d$ *is continuous, and for* $0 \leq j \leq m$, *the following two statements hold.*

(i) Uniform boundedness. *The partial derivatives of* $(t, x, \lambda) \mapsto f(t, x, \lambda)$ *with respect to* (x, λ), *given by* $D_{(2,3)}^j f : \mathcal{D} \times \Lambda \to L_j(\mathbb{R}^d \times Y, \mathbb{R}^d)$, *exist as continuous functions, and for all bounded sets* $B \subseteq \mathbb{R}^d$, *one has*

$$\sup_{t \in \mathbb{R}} \sup_{x \in B \cap \mathcal{D}(t)} \left| D_{(2,3)}^j f(t, x, \lambda) \right| < \infty \quad \text{for all } \lambda \in \Lambda.$$

(ii) Uniform continuity. *For all* $\lambda_0 \in \Lambda$ *and* $\varepsilon > 0$, *there exists a* $\delta > 0$ *such that*

---

[2] Readers unfamiliar with Banach spaces may consider an open subset of $\mathbb{R}^p$ instead of Y.

$$\sup_{\substack{t\in\mathbb{R}}} \sup_{\substack{x,y\in\mathcal{D}(t):\\ |x-y|<\delta}} \sup_{\substack{\lambda\in B_\delta(\lambda_0)}} \left|D^j_{(2,3)}f(t,x,\lambda) - D^j_{(2,3)}f(t,y,\lambda_0)\right| < \varepsilon.$$

Let $\lambda^* \in \Lambda$ be a parameter value and $\phi^* : \mathbb{R} \to \mathbb{R}^d$ be a bounded solution to (D_{λ^*}). Consider the dichotomy spectrum $\Sigma(\phi^*)$ of the variational equation

$$\dot{x} = D_2 f(t, \phi^*(t), \lambda^*)x \qquad\qquad (V_{\lambda^*})$$

having the transition matrix $\Phi(t,s)$.

Our interest concerns the behaviour of and near the reference solution ϕ^* when the parameter $\lambda \in \Lambda$ is varied close to λ^*. The following theorem from [190, Theorem 3.8] says that, provided the bounded reference solution ϕ^* is hyperbolic, the solution ϕ^* allows a unique smooth continuation near λ^*. We denote by $BC(\mathbb{R}^d)$ the linear space of bounded continuous functions from $\mathbb{R}$ to $\mathbb{R}^d$.

Theorem 2.3.2 (Continuation of Entire Solutions) *Consider the differential equation* (D_λ) *satisfying Hypothesis 2.3.1, and assume that*

$$0 \notin \Sigma(\phi^*), \qquad\qquad (2.2)$$

i.e. the entire bounded solution $\phi^* : \mathbb{R} \to \mathbb{R}^d$ *of* (D_{λ^*}) *is hyperbolic. Then there exist* $\rho, \varepsilon > 0$ *and a unique* C^m-*function*

$$\phi : B_\rho(\lambda^*) \to B_\varepsilon(\phi^*) \subset BC(\mathbb{R}^d)$$

such that $\phi(\lambda^*) = \phi^*$ *and every* $\phi(\lambda)$ *is a bounded and hyperbolic entire solution of* (D_λ) *with the same Morse index as* ϕ^*. *Here, the* Morse index *of the solution* ϕ^* *is the constant dimension of* $N(P(t))$, $t \in \mathbb{R}$, *where* P *denotes the invariant projector of the exponential dichotomy of* (V_{λ^*}).

The next remark addresses how Theorem 2.3.2 applies to autonomous systems.

Remark 2.3.3 (Nonautonomous Perturbations of Autonomous Equations) *It is well known that if* x^* *is an equilibrium to* $\dot{x} = g(x, \lambda^*)$ *with* $0 \notin \sigma(D_1 g(x^*, \lambda^*))$, *then* x^* *can be uniquely continued as an equilibrium in the parameter* λ. *However, the general setting of Theorem 2.3.2, where the parameter space* Λ *is a subset of a Banach space* Y, *allows to perturb the autonomous right hand side* g *by a possibly nonautonomous right hand side staying close to* g *or the constant parameter* λ^* *by small time-varying functions. By doing so, the result implies that hyperbolic fixed points* x^* *persist as entire bounded solutions under* L^∞-*small nonautonomous perturbations. Note that Theorem 2.3.2 does not apply to (nonconstant) periodic solutions* ϕ^* *of autonomous equations, since the existence of a Floquet multiplier* 1 *violates the hyperbolicity condition (2.2). Therefore, we refer to for example [96, p. 226, Theorem 4.1] for an appropriate persistence result.*

Due to the C^m-dependence of the perturbed solution $\phi(\lambda) \in BC(\mathbb{R}^d)$ on the parameter λ, one can approximate $\phi(\lambda)$ using a finite Taylor series in λ. In order to describe this, we have to introduce the following notations: given $j, n \in \mathbb{N}$, we write

$$P_j^<(n) := \left\{ (N_1, \ldots, N_j) \middle| \begin{array}{l} N_i \subseteq \{1, \ldots, n\} \text{ and } N_i \neq \emptyset \text{ for } 1 \leq i \leq j, \\ N_1 \cup \ldots \cup N_j = \{1, \ldots, n\}, \\ N_i \cap N_k = \emptyset \text{ for } i \neq k, \; i, k \in \{1, \ldots, j\}, \\ \max N_i < \max N_{i+1} \text{ for } 1 \leq i < j \end{array} \right\}$$

for the set of *ordered partitions* of $\{1, \ldots, n\}$ with *length* j and $\#N$ for the cardinality of a finite set $N \subset \mathbb{N}$. In case $N = \{n_1, \ldots, n_k\} \subseteq \{1, \ldots, n\}$ for $k \in \mathbb{N}$, $k \leq n$, we abbreviate $D^k g(x) x_N := D^k g(x) x_{n_1} \cdots x_{n_k}$ and

$$D^k g(x) x_1^{(k)} := D^k g(x) \underbrace{x_1 \cdots x_1}_{k \text{ times}}$$

for vectors $x, x_1 \in \mathbb{R}^d$. Here, the mapping $g : \mathbb{R}^d \to \mathbb{R}^d$ is assumed to be n-times continuously differentiable with derivatives $D^k g(x) \in L_k(\mathbb{R}^d)$.[3]

Using Taylor's theorem [229, p. 148, Theorem 4.A], we obtain

$$\phi(\lambda) = \phi^* + \sum_{j=1}^{m} \frac{1}{j!} D^j \phi(\lambda^*)(\lambda - \lambda^*)^{(j)} + R_m(\lambda) \tag{2.3}$$

with the coefficients $D^j \phi(\lambda^*) \in L_j(Y, BC(\mathbb{R}^d))$ and a remainder term R_m satisfying $\lim_{\lambda \to 0} \frac{1}{|\lambda|^m} R_m(\lambda) = 0$.

In the following, let us conveniently write $\phi(t, \lambda) := \phi_\lambda(t) := \phi(\lambda)(t) \in \mathbb{R}^d$. For $j \in \{1, \ldots, m\}$, we apply the higher order chain rule (see [139, p. 119, Lemma 6.16] for a reference in our notation) to the solution identity

$$D_1 \phi(t, \lambda) \equiv f(t, \phi_\lambda(t), \lambda) \quad \text{on } \Lambda_0 \text{ and all } t \in \mathbb{R}.$$

For $y_1, \ldots, y_n \in Y$, this yields the relation

$$D_2^n D_1 \phi(t, \lambda) y_1 \cdots y_n = D_2 f(t, \phi_\lambda(t), \lambda) D^n \phi(\lambda)(t) y_1 \cdots y_n$$

$$+ \sum_{j=2}^{n} \sum_{(N_1, \ldots, N_j) \in P_j^<(n)} D_{(2,3)}^j f(t, \phi_\lambda(t), \lambda) g^{\#N_1}(t, \lambda) y_{N_1} \cdots g^{\#N_j}(t, \lambda) y_{N_j},$$

where we abbreviate $g^{\#N_1}(t, \lambda) := \frac{d^{\#N_1}(\phi(t, \lambda), \lambda)}{d\lambda^{\#N_1}}$. Setting $\lambda = \lambda^*$ in this relation yields that the Taylor coefficients $D^n \phi(\lambda^*) \in L_n(Y, BC(\mathbb{R}^d)) \cong BC(L_n(Y, \mathbb{R}^d))$ fulfil the linearly inhomogeneous equation

$$\dot{X} = D_2 f(t, \phi^*(t), \lambda^*) X + H_n(t) \tag{I_n}$$

in $L_n(Y, \mathbb{R}^d)$, where the inhomogeneity $H_n : \mathbb{R} \to L_n(Y, \mathbb{R}^d)$ reads as

$$H_n(t) y_1 \cdots y_n :=$$

---

[3] We abbreviate $L_k(X, Y)$ for the space of bounded (symmetric) k-linear mappings from X^k into Y and $L_k(X) := L_k(X, X)$, $L_0(X, Y) := Y$.

$$\sum_{j=2}^{n} \sum_{(N_1,\ldots,N_j)\in P_j^{<}(n)} D_{(2,3)}^j f(t,\phi^*(t),\lambda^*) g^{\#N_1}(t,\lambda^*) y_{N_1} \cdots g^{\#N_j}(t,\lambda^*) y_{N_j}$$

and in particular $H_1(t) = D_3 f(t,\phi^*(t),\lambda^*)$.

Here, a phenomenon typical for the nonautonomous theory occurs: algebraic problems in an autonomous setting become dynamical problems. This means that instead of solving linear algebraic equations, one has to recursively find bounded solutions of a linear nonautonomous equation, in order to obtain the Taylor coefficients. We make a similar observation in Sect. 4.1 when dealing with nonautonomous invariant manifolds.

These preparations lead to the next result adapted from [190, Corollary 2.20].

Proposition 2.3.4 (Taylor Expansions of Continued Entire Solutions) *We con-sider the situation of Theorem 2.3.2, where the hyperbolic bounded entire solution ϕ^* of (D_{λ^*}) was continued via a unique C^m-function $\phi : B_\rho(\lambda^*) \to B_\varepsilon(\phi^*) \subset BC(\mathbb{R}^d)$. Then the coefficients $D^n\phi(\lambda^*) : \mathbb{R} \to L_n(Y, BC(\mathbb{R}^d))$ in the Taylor expansion (2.3) can be determined recursively from the* Lyapunov–Perron *integrals*

$$D^n\phi(\lambda^*) = \int_{-\infty}^{\infty} \Gamma_{\lambda^*}(\cdot,s) H_n(s)\,\mathrm{d}s \quad \text{for all } n \in \{1,\ldots,m\},$$

where Green's *function for (V_{λ^*}) is given by*

$$\Gamma_{\lambda^*}(t,s) := \begin{cases} \Phi(t,s)P(s) & : \ t \geq s, \\ -\Phi(t,s)(\mathrm{id} - P(s)) & : \ t < s \end{cases}$$

and $P : \mathbb{R} \to \mathbb{R}^{d\times d}$ is the invariant projector associated with the exponential dichotomy of (V_{λ^*}).

In the linear system (1.10) from Example 1.2.1, the continuation $\phi(\lambda) \in BC(\mathbb{R})$ is explicitly known. The next example concerns a perturbed logistic equation.

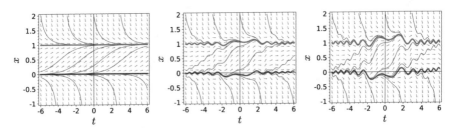

Figure 2.3 Persistence of bounded entire solutions from Theorem 2.3.2: Solution curves (blue) and nonautonomously perturbed equilibria of the logistic equation from Example 2.3.5 with coefficients $a(t) = \sin t^2$ and $\lambda = 0$ (left), $\lambda = 0.25$ (centre), $\lambda = 0.5$ (right) and the hyperbolic bounded entire solutions ϕ_λ (red and, respectively, green)

Example 2.3.5 (Nonautonomous Logistic Equation) *Let* $a : \mathbb{R} \to \mathbb{R}$ *be continuous. For* $\lambda = \lambda^* = 0$, *the* logistic equation

$$\dot{x} = x(1 - x) + \lambda a(t)$$

has an explicitly known family of bounded entire solutions, namely

$$\varphi_0(t, \tau, \xi) = \frac{e^{t-\tau}\xi}{1 + \xi(e^{t-\tau} - 1)} \quad \text{for all } \tau \in \mathbb{R} \text{ and } \xi \in [0, 1].$$

For initial values $\xi \in \{0, 1\}$, *they reduce to the equilibria of* $\dot{x} = x(1 - x)$, *while the solutions* $\varphi_0(\cdot, \tau, \xi)$ *are strictly increasing heteroclinics between* 0 *and* 1 *whenever* $\xi \in (0, 1)$. *Since the variational equation*

$$\dot{x} = (1 - 2\varphi_0(t, \tau, \xi))x = \frac{(\xi - 1)e^{\tau - t} + \xi}{(\xi - 1)e^{\tau - t} - \xi}x$$

is asymptotically autonomous, its dichotomy spectrum is given by

$$\Sigma(0) = \begin{cases} \{1\} & : \quad \xi = 0, \\ [-1, 1] & : \quad \xi \in (0, 1), \\ \{-1\} & : \quad \xi = 1 \end{cases} \quad \text{for all } \tau \in \mathbb{R},$$

and consequently, Theorem 2.3.2 applies to the hyperbolic equilibria 0 *and* 1, *but not to the nonhyperbolic homoclinics* $\varphi_0(\cdot, \tau, \xi)$ *when* $\xi \in (0, 1)$. *The unstable equilibrium* $\xi = 0$ *persists as an entire bounded solution* ϕ^1 *(see Fig. 2.3 (red)), whose Taylor coefficients in (2.3) read as*

$$D_2^j \phi^1(t, \lambda^*) = -\int_t^\infty e^{t-s} H_n(s)\, ds \quad \text{for all } j \in \mathbb{N}$$

with recursively approachable inhomogeneities

$$H_1(t) = a(t), \qquad\qquad H_2(t) = -2a(t)^2,$$
$$H_3(t) = -6a(t)H_2(t), \qquad H_4(t) = -6H_2(t)^2 - 8a(t)H_3(t)$$

and

$$H_5(t) = -20H_2(t)H_3(t) - 10a(t)H_4(t),$$
$$H_6(t) = -20H_3(t)^2 - 30H_2(t)H_4(t) - 12a(t)H_5(t).$$

The uniformly asymptotically stable equilibrium $\xi = 1$ *persists as bounded entire solution* ϕ^2 *(see Fig. 2.3 (green)) with, given* H_j *as above, Taylor coefficients*

$$D_2^j \phi^2(t, \lambda^*) = \int_{-\infty}^t e^{s-t} H_j(s)\, ds \quad \text{for all } j \in \mathbb{N}.$$

Finally, in order to provide some perspective, the situation drastically changes on half lines $\mathbb{R}_0^+$ and $\mathbb{R}_0^-$, since hyperbolic solutions are embedded into a family of bounded solutions forming the stable/unstable manifold and are not isolated. For more detailed comments, we refer to the discrete time case in Sect. 5.3.2.

Remarks

We first note that general introductions to exponential dichotomies are provided in [54, 59].

Dynamical Spectra The *Sacker–Sell spectrum* was introduced in [210] for linear skew product flows over a compact base space. In contrast, we directly work with linear differential equations (L) and hence prefer the terminology *dichotomy spectrum* [201, 215]. An alternative approach via the *Morse spectrum* can be found in [202]. While the dichotomy spectrum is based on the entire system (L) (i.e. its transition matrix), Lyapunov exponents [1, 24, 53] and the related spectrum rely on individual solutions. A solution-based spectral notion yielding uniform exponential growth is the *Bohl spectrum* from [59, p. 174] and [68].

The counterpart of the Spectral Theorem 2.1.6 based on Lyapunov exponents is the deterministic *multiplicative ergodic theorem*, which can be found in [7, 123].

Both spectral intervals and Lyapunov exponents measure exponential growth of solutions, but we do not want to conceal that they fail to indicate rotational behaviour, which is a crucial ingredient for, e.g. Hopf bifurcations. A suitable concept for this purpose might be the *rotation number* [119, 171].

The numerical approximation of nonautonomous spectra became popular over the last 20 years. Adequate techniques were developed in [62–64], as well as in [74] for the rotational number.

Stability The famous example of an unstable solution whose linearisation has negative Lyapunov exponents is due to Perron [183, § 2]. While uniform exponential stability can be characterised in terms of the dichotomy spectrum (Proposition 2.2.1(a)), we are not aware of such necessary and sufficient descriptions for exponential stability based on the Lyapunov spectrum. Nevertheless, a sufficient condition for exponential stability is given in Theorem A.1.3 (see also [24, p. 71, Theorem 3.9] or [153, p. 55ff, Chapter 4]). Concerning criteria for instability, see [153, pp. 61ff, Chapter 7].

Continuation The proof of Theorem 2.3.2 is essentially based on the implicit function theorem. This has two consequences. First, using a quantitative version of this result (for instance, [103]), one can obtain estimates for the size of the neighbourhoods occurring. This, in turn, yields robustness results on the magnitude of parametric perturbations. Such issues are also addressed in [34]. Under rather strong assumptions, the local solution branches from Theorem 2.3.2 can be continued globally, but this involves a solid topological machinery [196]. Second, Theorem 2.3.2 can be refined to continue solutions in subspaces $X_0 \subseteq X$

of $BC(\mathbb{R}^d)$, like for example the almost periodic functions, provided *admissibility* conditions hold: for each inhomogeneity $h \in X$, there exists a unique solution to $\dot{x} = D_2 f(t, \phi^*(t), \lambda^*)x + h(t)$ in a function space X_0.

Besides Taylor approximation, an alternative method to compute (branches of) bounded entire solutions is given in [127].

Chapter 3
Nonautonomous Bifurcation

In parameter-dependent autonomous systems, a bifurcation indicates a topological change of the phase portrait [148]. Such a change may involve the creation or disappearance of certain solutions, indicating a change of stability or a transition in the structure and shape of an attractor. While the notion of topological equivalence is somewhat problematic in a nonautonomous setting, the other above properties indicating a bifurcation are more accessible for generalisations. Nonetheless, it is a conceptional problem in nonautonomous bifurcation theory to identify a suitable class of bifurcating objects.

From a stability perspective, one might understand a bifurcation as a change in the structure of attracting or repelling sets, which forms the basis for *attractor bifurcation* (see [199, 200] or [163, pp. 114ff]). Before discussing several results, we first introduce and exemplify various forms of attraction and repulsion.

From a continuation perspective, there is the intuition that bifurcating objects should reflect the temporal driving (quasi- or almost periodic, bounded, etc.). This led to *solution bifurcation*, which can be approached using dynamical systems concepts [150, 152] or analytical machinery [187]. In terms of the dichotomy spectrum, a necessary condition for such bifurcations can be given, but as a matter of course we formulate sufficient ones.

In a skew product setting, it is natural to look for bifurcating invariant graphs or minimal sets [173]. A benefit in this framework is that the bifurcating objects directly inherit properties of the driving system.

All these approaches yield not totally independent scenarios indicating that nonhyperbolicity is a significantly wider concept in the nonautonomous theory and allows a wide range of different behaviours. Among them is the feature that nonautonomous bifurcations appear as a two-step transition [121]. This phenomenon was originally observed by L. Arnold and his co-workers in a random setting. Transferred to deterministic problems, one aspect is that in stability changes first uniformity is lost, before eventually an unstable regime arises. In terms of the dichotomy spectrum, this reflects the fact that a dominant (non-singleton) spectral interval contains the stability boundary for a continuum of parameters (see Example 1.2.4). As

V. Anagnostopoulou et al., *Nonautonomous Bifurcation Theory*, Frontiers in Applied Dynamical Systems: Reviews and Tutorials 10, https://doi.org/10.1007/978-3-031-29842-4_3

additional aspect we point out that during the critical regime very complicated dynamics occurs [174].

This chapter presents prototypical results describing nonautonomous bifurcations and transitions. They include attractor and solution bifurcations, as well as bifurcations of minimal sets. In addition, we compare these approaches.

3.1 Attractor Bifurcation

A simple example for a bifurcation of an attractor was discussed in Example 1.2.3. In the following, a general bifurcation pattern will be derived, which ensures that under certain conditions on the Taylor coefficients of the right hand side f in (D_λ), an attractor changes qualitatively under variation of the parameter. This allows us to extend autonomous bifurcation patterns of transcritical and pitchfork type. Although the attractor discussed in Example 1.2.3 was a *global* attractor, the bifurcation results presented here only yield properties for *local* attractors. Thus, we first need to introduce notions of local attractiveness (and repulsiveness) for processes [199]. In order to describe repulsiveness, it is convenient to consider (D) under time reversal. This leads to the *time-reversed differential equation*

$$\dot{x} = -f(-t, x),$$

and we denote the corresponding process by φ^-.

Definition 3.1.1 (Local Attractors and Repellers for Processes) *Let* $\mathcal{A}, \mathcal{R} \subset \mathcal{D}$ *be compact and invariant nonautonomous sets.*

   (i) $\mathcal{A}$ *is called* local attractor *on* $\mathbb{I} = \mathbb{R}$ *if there exists an* $\eta > 0$ *with*

$$\lim_{t \to \infty} \sup_{\tau \in \mathbb{R}} d\big(\varphi(t + \tau, \tau, B_\eta(\mathcal{A}(\tau))), \mathcal{A}(t + \tau)\big) = 0.$$

The supremum of all positive η *with this property is denoted by* $\rho(\mathcal{A})$ *and called* attraction radius *of* $\mathcal{A}$ *on* $\mathbb{I} = \mathbb{R}$.

   (ii) $\mathcal{A}$ *is called* local attractor *on* $\mathbb{I} = \mathbb{R}_0^-$ *if there exists an* $\eta > 0$ *with*

$$\lim_{t \to \infty} d\big(\varphi(\tau, \tau - t, B_\eta(\mathcal{A}(\tau - t))), \mathcal{A}(\tau)\big) = 0 \quad \text{for all } \tau \in \mathbb{I}. \tag{3.1}$$

The supremum of all positive η *with this property is denoted by* $\rho(\mathcal{A})$ *and called* attraction radius *of* $\mathcal{A}$ *on* $\mathbb{I} = \mathbb{R}_0^-$.

   (iii) $\mathcal{A}$ *is called* local attractor *on* $\mathbb{I} = \mathbb{R}_0^+$ *if there exists an* $\eta > 0$ *with*

$$\lim_{t \to \infty} d\big(\varphi(\tau + t, \tau, B_\eta(\mathcal{A}(\tau))), \mathcal{A}(\tau + t)\big) = 0 \quad \text{for all } \tau \in \mathbb{I}.$$

The supremum of all $\eta > 0$ *such that there exists a* $\tilde{\tau} \geq 0$ *with*

$$\lim_{t\to\infty} d\big(\varphi(\tau + t, \tau, B_\eta(\mathcal{A}(\tau))), \mathcal{A}(\tau + t)\big) = 0 \quad \text{for all } \tau \geq \tilde{\tau}$$

is denoted by $\rho(\mathcal{A})$ *and called* attraction radius *of* $\mathcal{A}$ *on* $\mathbb{I} = \mathbb{R}_0^+$.

(iv) $\mathcal{R}$ *is called* local repeller *on* $\mathbb{I} = \mathbb{R}, \mathbb{R}_0^-, \mathbb{R}_0^+$ *if* $\mathcal{R}^- := \{(t, x) \in \mathbb{R} \times \mathbb{R}^d :$ $(-t, x) \in \mathcal{R}\}$ *is a local attractor on* $\mathbb{I} = \mathbb{R}, \mathbb{R}_0^+, \mathbb{R}_0^-$ *(with respect to* φ^-*), respectively. Note here that the roles of the half lines* $\mathbb{R}_0^+$ *and* $\mathbb{R}_0^-$ *have changed due to time reversal.*

In addition, we call a solution $\phi : \mathbb{I} \to \mathbb{R}^d$ *locally attracting on* $\mathbb{I}$ *if it is a local attractor on* $\mathbb{I}$*, interpreted as a nonautonomous set, and a solution* $\phi : \mathbb{I} \to \mathbb{R}^d$ *is called* locally repulsive *on* $\mathbb{I}$ *if it is a local repeller on* $\mathbb{I}$.

Note that a local attractor on $\mathbb{R}_0^-$ is also frequently called *local pullback attractor*, while a local attractor on $\mathbb{R}_0^+$ is also called *local forward attractor* [139]. In addition, a local repeller on $\mathbb{R}_0^+$ is called *local pullback repeller*, while a local repeller on $\mathbb{R}_0^-$ is called *local forward repeller*.

We illustrate the above notions by means of a simple example.

Example 3.1.2 (Nonautonomous Bernoulli Equation) *We consider the nonautonomous Bernoulli equation*

$$\dot{x} = a(t)x - b(t)x^3,$$

where we assume that $a : \mathbb{R} \to \mathbb{R}$ *and* $b : \mathbb{R} \to (\beta, \infty)$ *are continuous functions for some* $\beta > 0$*. We note that this differential equation generalises (1.12) from Example 1.2.3, and as in that example, it is possible to find an explicit representation for the associated process [152, Equation (5.3)]. Using this representation, it is easy to conclude that the trivial solution is a*

(i) *Local forward attractor if* $\limsup_{t\to\infty} a(t) \leq 0$

(ii) *Local pullback attractor if* $\limsup_{t\to-\infty} a(t) \leq 0$

(iii) *Local attractor on* $\mathbb{R}$ *if* $a(t) \leq 0$ *for all* $t \in \mathbb{R}$

(iv) *Local forward repeller if* $\limsup_{t\to-\infty} \frac{a(t)}{b(t)} < 0$

(v) *Local pullback repeller if* $\limsup_{t\to\infty} \frac{a(t)}{b(t)} < 0$

(vi) *Local repeller on* $\mathbb{R}$ *if* $\sup_{t\in\mathbb{R}} \frac{a(t)}{b(t)} < 0$

These conditions confirm that a local forward attractor and a local pullback repeller concern the dynamical behaviour for $t \to \infty$*, while a local pullback attractor and a local forward repeller relate to the limit* $t \to -\infty$.

The following results in this section are due to [199, 200] and formulated for scalar differential equations, but by means of the nonautonomous centre manifold reduction presented in Sect. 4.1, they can be extended to higher-dimensional problems.

Let $\Lambda \subseteq \mathbb{R}$ be an interval and $\mathbb{I} \subseteq \mathbb{R}$ be an unbounded interval. We now restrict to one-dimensional nonautonomous differential equations (D_λ) with right hand side $f : \mathcal{D} \times \Lambda \to \mathbb{R}$, where $\mathcal{D} \subset \mathbb{R} \times \mathbb{R}$. Let us assume that (D_λ) possesses a continuously parametrised family $\phi_\lambda, \lambda \in \Lambda$, of bounded entire solutions, i.e.

$$\dot{\phi}_\lambda(t) \equiv f(t, \phi_\lambda(t), \lambda) \quad \text{for all } t \in \mathbb{R}.$$

Since we are interested in bifurcations of attractors close to this family of reference solutions, we consider the scalar variational equation

$$\dot{x} = D_2 f(t, \phi_\lambda(t), \lambda)x \qquad\qquad (Y_\lambda)$$

with transition mapping $\Phi_\lambda(t, s) = \exp\left(\int_s^t D_2 f(r, \phi_\lambda(r), \lambda)\, dr\right)$. The next condition postulates that (Y_λ) changes stability transversally at a parameter value λ^*.

Hypothesis 3.1.3 (Transversal Exchange of Stability) *We assume that there exist $\lambda^* \in \Lambda$, $K \geq 1$ and functions $\gamma_+, \gamma_- : \Lambda \to \mathbb{R}$ that are either both increasing or both decreasing such that* $\lim_{\lambda \to \lambda^*} \gamma_+(\lambda) = \lim_{\lambda \to \lambda^*} \gamma_-(\lambda) = 0$, *and*

$$\Phi_\lambda(t, s) \leq K e^{\gamma_+(\lambda)(t-s)}, \quad \Phi_\lambda(s, t) \leq K e^{\gamma_-(\lambda)(s-t)} \quad \text{for all } s \leq t, \lambda \in \Lambda.$$

Note that this hypothesis ensures the inclusion

$$\Sigma_{\mathbb{I}}(\lambda) \subseteq \left[\min\{\gamma_+(\lambda), \gamma_-(\lambda)\}, \max\{\gamma_+(\lambda), \gamma_-(\lambda)\}\right]$$

for the dichotomy spectrum $\Sigma_{\mathbb{I}}(\lambda)$ of (Y_λ), and at the critical value (cf. Fig. 3.1)

$$\lim_{\lambda \to \lambda^*} d(\Sigma(\lambda), \{0\}) = 0.$$

Under appropriate assumptions on the nonlinearities, we obtain nonautonomous counterparts to both the transcritical and pitchfork bifurcation patterns. Let us first deal with a transcritical bifurcation scenario [200, Theorem 5.1]. It guarantees that attraction radii of attractive solutions vanish at the critical parameter value.

Theorem 3.1.4 (Transcritical Attractor Bifurcation) *We suppose that Hypothesis 3.1.3 holds and (D_λ) is of class C^3. Assume that*

$$-\infty < \liminf_{\lambda \to \lambda^*} \inf_{t \in \mathbb{I}} D_2^2 f(t, \phi_\lambda(t), \lambda) \leq \limsup_{\lambda \to \lambda^*} \sup_{t \in \mathbb{I}} D_2^2 f(t, \phi_\lambda(t), \lambda) < 0 \quad (3.2)$$

is satisfied and the third-order term fulfils

Figure 3.1 Dichotomy spectra $\Sigma_{\mathbb{I}}(\lambda)$ under Hypothesis 3.1.3 degenerating to the singleton $\{0\}$ in the limit $\lambda \to \lambda^*$ for **(a)** increasing functions $\gamma_\pm$ and **(b)** decreasing functions $\gamma_\pm$

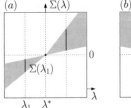
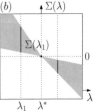

$$\lim_{x \to 0} \sup_{\lambda \in (\lambda^* - |x|, \lambda^* + |x|)} \sup_{t \in \mathbb{I}} x \int_0^1 (1 - \vartheta)^2 D_2^3 f(t, \phi_\lambda(t) + \vartheta x, \lambda) \, d\vartheta = 0,$$

$$\limsup_{\lambda \to \lambda^*} \limsup_{x \to 0} \sup_{t \in \mathbb{I}} \frac{K x^2 \int_0^1 (1 - \vartheta)^2 D_2^3 f(t, \phi_\lambda(t) + \vartheta x, \lambda) \, d\vartheta}{1 - \min\{\gamma_+(\lambda), -\gamma_-(\lambda)\}} < 1.$$

Then there exist $\lambda_- < \lambda^* < \lambda_+$ so that the following statements hold:

(a) *For increasing functions γ_+, γ_-, the solution ϕ_λ is locally attracting on $\mathbb{I}$ for $\lambda \in (\lambda_-, \lambda^*)$. At $\lambda = \lambda^*$, the equation (D_λ) admits an attractor bifurcation in the sense that*

$$\lim_{\lambda \nearrow \lambda^*} \rho(\phi_\lambda) = 0.$$

(b) *For decreasing functions γ_+, γ_-, the solution ϕ_λ is locally attracting on $\mathbb{I}$ for $\lambda \in (\lambda^*, \lambda_+)$. At $\lambda = \lambda^*$, the equation (D_λ) admits an attractor bifurcation in the sense that*

$$\lim_{\lambda \searrow \lambda^*} \rho(\phi_\lambda) = 0.$$

Note that the statement in [200, Theorem 5.1] deviates from the above formulation; for a proof how to translate both versions, see [188, Theorem 4.1].

The formulation of Theorem 3.1.4 does not mention that ϕ_λ is locally repelling for $\lambda \in (\lambda^*, \lambda_+)$ in case (i) and for $\lambda \in (\lambda_-, \lambda^*)$ in case (ii), and the corresponding repulsion radii bifurcate, see [200, Theorem 5.1] for definitions and details. In this reference also dual assertions are stated for the case when (3.2) is replaced by

$$0 < \liminf_{\lambda \to \lambda^*} \inf_{t \in \mathbb{I}} D_2^2 f(t, \phi_\lambda(t), \lambda) \leq \limsup_{\lambda \to \lambda^*} \sup_{t \in \mathbb{I}} D_2^2 f(t, \phi_\lambda(t), \lambda) < \infty.$$

Example 3.1.5 (Nonautonomous Logistic Equation) *Let the continuous functions $a, b : \mathbb{I} \to (0, \infty)$ be bounded and bounded away from zero. We consider the trivial solution $\phi_\lambda(t) \equiv 0$ of the logistic equation*

$$\dot{x} = \lambda a(t) x - b(t) x^2. \tag{3.3}$$

Let $\lambda^* = 0$. Since the function b is bounded and bounded away from 0, the condition (3.2) holds, and the subsequent two conditions on the third-order terms are also fulfilled, since (3.3) has only terms up to order 2 in x. It is also clear that equation (3.3) satisfies Hypothesis 3.1.3 with increasing functions γ_+ and γ_-. Hence, (3.3) admits a transcritical attractor bifurcation at $\lambda^* = 0$: the trivial solution ϕ_λ is locally attracting for $\lambda < 0$, and we have $\lim_{\lambda \nearrow 0} \rho(\mathcal{A}_{\phi_\lambda}) = 0$. It is easy to obtain the following estimates for the attraction radius $\rho(\mathcal{A}_{\phi_\lambda})$ for $\lambda < 0$:

$$\rho(\mathcal{A}_{\phi_\lambda}) \in \begin{cases} \left[-\sup_{t \in \mathbb{R}} \lambda \frac{a(t)}{b(t)}, -\inf_{t \in \mathbb{R}} \lambda \frac{a(t)}{b(t)} \right] & : \quad \mathbb{I} = \mathbb{R}, \\[2mm] \left[-\limsup_{t \to -\infty} \lambda \frac{a(t)}{b(t)}, -\liminf_{t \to -\infty} \lambda \frac{a(t)}{b(t)} \right] & : \quad \mathbb{I} = \mathbb{R}_0^-, \\[2mm] \left[-\limsup_{t \to \infty} \lambda \frac{a(t)}{b(t)}, -\liminf_{t \to \infty} \lambda \frac{a(t)}{b(t)} \right] & : \quad \mathbb{I} = \mathbb{R}_0^+. \end{cases}$$

An illustration is given in Fig. 3.2.

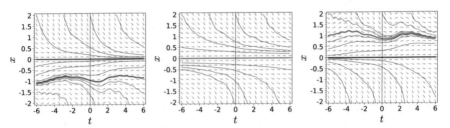

Figure 3.2 Transcritical bifurcation as in Theorem 3.1.4(a): solution curves (blue) of the logistic equation (3.3) with coefficients $a(t) = 1 + \frac{\sin t}{2}$ and $b(t) = \frac{1}{4} + \frac{\cos t^2}{8}$ for $\lambda = -\frac{1}{4}$ (left, 0 is asymptotically stable), $\lambda = 0$ (centre, 0 is semi-stable) and $\lambda = \frac{1}{4}$ (right, 0 is repulsive)

A similar bifurcation pattern is given by the nonautonomous pitchfork attractor bifurcation [200, Theorem 6.1]. In contrast to the transcritical bifurcation, the attracting object is a nontrivial attractor that becomes trivial at the bifurcation point.

Theorem 3.1.6 (Pitchfork Attractor Bifurcation) *Suppose that Hypothesis 3.1.3 holds and (D_λ) is of class C^4 with*

$$D_2^2 f(t, \phi_\lambda(t), \lambda) = 0 \quad \text{for all } t \in \mathbb{I} \text{ and } \lambda \in \Lambda.$$

Furthermore, we assume that there exists a $\lambda^* \in \Lambda$ such that the following hypotheses hold:

(i) If the functions γ_+ and γ_- are increasing, then we assume that

$$-\infty < \liminf_{\lambda \to \lambda^*} \inf_{t \in \mathbb{I}} D_2^3 f(t, \phi_\lambda(t), \lambda) \le \limsup_{\lambda \to \lambda^*} \sup_{t \in \mathbb{I}} D_2^3 f(t, \phi_\lambda(t), \lambda) < 0.$$

(ii) If the functions γ_+ and γ_- are decreasing, then we assume that

$$0 < \liminf_{\lambda \to \lambda^*} \inf_{t \in \mathbb{I}} D_2^3 f(t, \phi_\lambda(t), \lambda) \le \limsup_{\lambda \to \lambda^*} \sup_{t \in \mathbb{I}} D_2^3 f(t, \phi_\lambda(t), \lambda) < \infty.$$

In addition, assume that

$$\lim_{x \to 0} \sup_{\lambda \in (\lambda^* - x^2, \lambda^* + x^2)} \sup_{t \in \mathbb{I}} x \int_0^1 (1-\vartheta)^3 D^4 f(t, \phi_\lambda(t) + \vartheta x, \lambda) \, d\vartheta = 0,$$

$$\limsup_{\lambda \to \lambda^*} \limsup_{x \to 0} \sup_{t \in \mathbb{I}} \frac{K x^3 \int_0^1 (1-\vartheta)^3 D^4 f(t, \phi_\lambda(t) + \vartheta x, \lambda) \, d\vartheta}{1 - \min\{\gamma_+(\lambda), -\gamma_-(\lambda)\}} < 3.$$

Then there exist $\lambda_- < \lambda^* < \lambda_+$ so that the following statements hold:

(a) For increasing functions γ_+, γ_-, the solution ϕ_λ, $\lambda \in (\lambda_-, \lambda^*)$, is a local attractor bifurcating into a nontrivial local attractor $\mathcal{A}_\lambda$, $\lambda \in (\lambda^*, \lambda_+)$, and

$$\lim_{\lambda \nearrow \lambda^*} \sup_{t \in \mathbb{I}} d(\mathcal{A}_\lambda(t), \{\phi_\lambda(t)\}) = 0.$$

(b) *For decreasing functions* γ_+, γ_-, *the solution* ϕ_λ, $\lambda \in (\lambda^*, \lambda_+)$, *is a local attractor bifurcating into a nontrivial local attractor* $\mathcal{A}_\lambda$, $\lambda \in (\lambda_-, \lambda^*)$, *and*

$$\lim_{\lambda \searrow \lambda^*} \sup_{t \in \mathbb{I}} d(\mathcal{A}_\lambda(t), \{\phi_\lambda(t)\}) = 0.$$

Similar to the situation in Theorem 3.1.4, the proof of this statement follows from [200, Theorem 6.1] and [188, Theorem 4.2]. A dual version to Theorem 3.1.6 for pitchfork bifurcations into nontrivial repellers was given in [200, Theorem 6.1].

3.2 Solution Bifurcation

The concept of a *solution bifurcation* is typical in abstract branching theory covered in [48, 132, 229] and understood as follows. We restrict to one-parameter bifurcations, where $\Lambda \subseteq \mathbb{R}$ is open and suppose that for a fixed parameter $\lambda^* \in \Lambda$, the equation (D_{λ^*}) possesses a bounded entire reference solution ϕ^*. Then one says that (D_λ) undergoes a *bifurcation* at $\lambda = \lambda^*$ along ϕ^*, if there exists a convergent parameter sequence $(\lambda_n)_{n \in \mathbb{N}}$ in Λ with limit λ^* so that each (D_{λ_n}) has at least two distinct bounded entire solutions ϕ_n and $\bar{\phi}_n$ with

$$\lim_{n \to \infty} \phi_n = \lim_{n \to \infty} \bar{\phi}_n = \phi^*.$$

In the following, these limits are understood in the supremum norm. Nevertheless, other topologies such as uniform convergence on compact subsets might be appropriate for a more general approach to solution bifurcations.

To search for bounded entire solutions is a natural choice when tackling nonautonomous equations (D_λ) using tools from functional analysis. For more specific time-dependencies in (D_λ), it might be suitable to consider e.g. convergent or almost periodic solutions. Also corresponding solutions on half lines are feasible. One speaks of a *subcritical* or a *supercritical bifurcation*, if $(\lambda_n)_{n \in \mathbb{N}}$ can be chosen according to $\lambda_n < \lambda^*$ or $\lambda_n > \lambda^*$, respectively (see Fig. 3.3).

This definition immediately yields a necessary condition for a bifurcation.

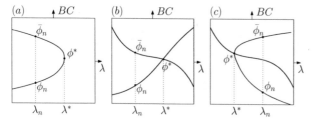

Figure 3.3 Diagram of a **(a)** subcritical, **(b)** transcritical and **(c)** supercritical solution bifurcation in the space $BC(\mathbb{R}^d)$ of bounded entire solutions to (D_λ)

Proposition 3.2.1 (Necessary Condition for Bifurcation) *Let* $\lambda^* \in \Lambda$. *If a bounded entire solution* ϕ^* *of* (D_{λ^*}) *bifurcates at* λ^*, *then* ϕ^* *is not hyperbolic on* $\mathbb{R}$.

Proof. If we suppose that $0 \notin \Sigma(\lambda^*)$, then Theorem 2.3.2 yields neighbourhoods $\Lambda_0 \subseteq \Lambda$ for λ^* and $U \subseteq BC(\mathbb{R}^d)$ for ϕ^*, so that (D_λ) has a unique entire solution $\phi_\lambda \in U$ for all parameters $\lambda \in \Lambda_0$. Hence, ϕ^* cannot bifurcate at λ^*. $\square$

A first type fitting this framework is the bifurcations of bounded entire solutions discussed in [187, 191]. In this setting, only unstable solutions of at least two-dimensional equations (D_λ) can bifurcate. We refer to the analogous discrete-time situation in Sect. 6.2 for a more detailed discussion.

The following subsections describe three further types of nonautonomous bifurcations related to solutions that bifurcate in a specific way. Although all types are fundamentally different, they satisfy the above necessary condition.

3.2.1 One-Dimensional Solution Bifurcations

The results of this subsection are due to J.A. Langa, J.C. Robinson and A. Suárez [150, 152] and focus on scalar parametrised differential equations (D_λ) having a smooth family of bounded entire solutions $\phi_\lambda : \mathbb{R} \to \mathbb{R}$, $\lambda \in \Lambda$, and $\phi^* = \phi_{\lambda^*}$. Here we assume that the j-th order derivatives $\phi_\lambda^j : \mathbb{R} \to \mathbb{R}$, $\phi_\lambda^j(t) := \frac{d^j}{d\lambda^j}\phi_\lambda(t)$ of ϕ_λ with respect to the parameter λ exist for $j \in \{1, 2\}$. Moreover, we assume that the partial derivatives $D_2^{j_1} D_3^{j_2} f : \mathcal{D} \to \mathbb{R}$ exist as continuous functions up to order $j_1 + j_2 \leq 3$.

Theorem 3.2.2 (Transcritical Solution Bifurcation) *Consider* (D_λ) *with a family of bounded entire solutions* $\phi_\lambda : \mathbb{R} \to \mathbb{R}$, *and suppose that for some* $\lambda^* \in \Lambda$ *the nonhyperbolicity condition*

$$D_2 f(t, \phi^*(t), \lambda^*) = 0 \quad \text{for all } t \in \mathbb{R}$$

and

$$\limsup_{t \to \pm\infty} D_2^2 f(t, \phi^*(t), \lambda^*) < 0,$$

$$-\infty < \liminf_{t \to \pm\infty} \left(\phi_{\lambda^*}^1(t) + \frac{D_2 D_3 f(t, \phi^*(t), \lambda^*)}{D_2^2 f(t, \phi^*(t), \lambda^*)} \right),$$

$$\limsup_{t \to \pm\infty} \left(\phi_{\lambda^*}^1(t) + \frac{D_2 D_3 f(t, \phi^*(t), \lambda^*)}{D_2^2 f(t, \phi^*(t), \lambda^*)} \right) < 0$$

hold. If there exists a function $h : \mathbb{R} \to \mathbb{R}_0^+$ *satisfying*

$$|\psi(t, \lambda)| \leq h(t), \qquad |D_2\gamma(t, x, \lambda)| \leq h(t), \qquad |D_3\gamma(t, x, \lambda)| \leq h(t)$$

for all $(t, x) \in \mathcal{D}$, λ *near* 0 *and* $-\infty < \liminf_{t \to \pm\infty} \frac{h(t)}{D_2^2 f(t, \phi^*(t), \lambda^*)}$ *with*

$$\psi(t, \lambda) := \int_0^1 (1-s)\big[D_2^3 f(t, \phi_{\lambda^*+s\lambda}(t), \lambda^* + s\lambda)\phi_{\lambda^*+s\lambda}^1(t)^2$$
$$+ 2D_3 D_2^2 f(t, \phi_{\lambda^*+s\lambda}(t), \lambda^* + s\lambda)\phi_{\lambda^*+s\lambda}^1(t)$$
$$+ D_2^2 f(t, \phi_{\lambda^*+s\lambda}(t), \lambda^* + s\lambda)\phi_{\lambda^*+s\lambda}^2(t)$$
$$+ D_2 D_3^2 f(t, \phi_{\lambda^*+s\lambda}(t), \lambda^* + s\lambda)\big]\, ds,$$
$$\gamma(t, x, \lambda) := \int_0^1 (\vartheta - 1) D_2^2 f(t, \phi_{\lambda^*+\lambda}(t) + \vartheta x, \lambda^* + \lambda)\, d\vartheta$$
$$+ \tfrac{1}{2} D_2^2 f(t, \phi^*(t), \lambda^*),$$

then the solution branch $(\phi_\lambda)_{\lambda \in \Lambda}$ *undergoes a* transcritical *bifurcation at* λ^* *in the following sense: there exist reals* $\lambda_- < \lambda^* < \lambda_+$ *in* Λ *such that*

(a) *For* $\lambda \in (\lambda_-, \lambda^*)$, *the solution* ϕ_λ *is locally pullback attracting, and there exists another solution* $\phi_\lambda^- : \mathbb{R}_0^+ \to \mathbb{R}$, *which is locally pullback repelling and satisfies*

$$\phi_\lambda^-(t) < \phi_\lambda(t), \qquad \lim_{\lambda \nearrow \lambda^*} \phi_\lambda^-(t) = \phi^*(t) \quad \text{for all } t \geq 0\,.$$

(b) ϕ^* *is neither locally pullback attracting nor locally pullback repelling.*

(c) *For* $\lambda \in (\lambda^*, \lambda_+)$, *the solution* ϕ_λ *is locally pullback repelling, and there exists another solution* $\phi_\lambda^+ : \mathbb{R}_0^- \to \mathbb{R}$, *which is locally pullback attracting, and satisfies*

$$\phi_\lambda(t) < \phi_\lambda^+(t), \qquad \lim_{\lambda \searrow \lambda^*} \phi_\lambda^+(t) = \phi^*(t) \quad \text{for all } t \leq 0\,.$$

Proof. Apply [152, Theorem 7] with

$$G(t, x, \lambda) := f(t, x + \phi_{\lambda^*+\lambda}(t), \lambda^* + \lambda) - f(t, \phi_{\lambda^*+\lambda}(t), \lambda^* + \lambda),$$

i.e. the equation of perturbed motion with respect to the solution ϕ_λ. The expressions for ψ, γ are due to Taylor's theorem with integral remainder [229, pp. 148–149, Theorem 4.A(b)]. $\square$

We now apply this theorem to the nonautonomous logistic equation from Example 3.1.5.

Example 3.2.3 (Nonautonomous Logistic Equation Revisited) *Consider again the nonautonomous* logistic equation

$$\dot{x} = \lambda a(t)x - b(t)x^2\,, \tag{3.4}$$

where $a, b : \mathbb{R} \to (0, \infty)$ *are continuous functions that are bounded and bounded away from zero. We choose the trivial solution* $\phi_\lambda(t) \equiv 0$ *and* $\lambda^* = 0$ *as critical* *parameter. We have seen in Example 3.1.5 that this differential equation admits a transcritical bifurcation in the sense of Sect. 3.1. It is straightforward to see that all assumptions of Theorem 3.2.2 are satisfied with $\psi(t, \lambda) \equiv \gamma(t, x, \lambda) \equiv 0$. Thus, (3.4) admits a transcritical solution bifurcation as well. Referring to Fig. 3.2, for $\lambda < 0$, the locally pullback repelling solution ϕ_λ^- below ϕ_λ is indicated in red, while for $\lambda > 0$, the locally pullback attracting solution ϕ_λ^+ above ϕ_λ is marked in green and explicitly given by $\phi_\lambda^+(t) = \left(\lambda^2 \int_{-\infty}^{t} a(s)b(s) \int_t^s a(r) \, dr \, ds \right)^{-1}$.*

Note that in addition to the transcritical bifurcation presented above, also fold bifurcations and pitchfork bifurcations have been treated similarly to Theorem 3.2.2 in [150, 152].

3.2.2 Shovel Bifurcation

Let us return to systems (D_λ) in $\mathbb{R}^d$. The phenomenon of a shovel bifurcation is based on the assumption that a reference solution, being unstable on the negative half line $\mathbb{R}_0^-$ while asymptotically stable on the positive half line $\mathbb{R}_0^+$, gains or loses stability as the bifurcation parameter moves through a critical value.

To describe such a scenario, we first assume that we have a family of solutions that satisfy the necessary condition for a bifurcation from Proposition 3.2.1 in the sense that zero is contained in a spectral interval.

Hypothesis 3.2.4 *Suppose that (D_λ) possesses a family of bounded entire solutions $\phi_\lambda : \mathbb{R} \to \mathbb{R}^d$ such that the dichotomy spectra of the variational equation*

$$\dot{x} = D_2 f(t, \phi_\lambda(t), \lambda) x$$

allow for all $\lambda \in \Lambda$ a splitting of the form (here, $\dot\cup$ denotes the disjoint union)

$$\Sigma(\lambda) = \Sigma_-(\lambda) \dot\cup \sigma(\lambda),$$
$$\Sigma^-(\lambda) = \Sigma_-^-(\lambda) \dot\cup \sigma^-(\lambda), \text{ and}$$
$$\Sigma^+(\lambda) = \Sigma_-^+(\lambda) \dot\cup \sigma^+(\lambda)$$

into dominant intervals $\sigma(\lambda)$, $\sigma^-(\lambda)$ and $\sigma^+(\lambda)$ and a remaining spectral part with $\sup_{\lambda \in \Lambda} \max \Sigma_-(\lambda) < 0$. Let m be the multiplicity of $\sigma(\lambda)$. We violate hyperbolicity by assuming that there exists a $\lambda^* \in \Lambda$ with (see Fig. 3.4)

$$\max \Sigma(\lambda^*) = \max \sigma(\lambda^*) = 0.$$

If for some $\lambda \in \Lambda$, we have $\max \Sigma_-(\lambda) < 0$, then the nonautonomous equation (D_λ) possesses a centre manifold [194, Theorem 3.2], and the stability analysis for the bounded entire solution ϕ_λ reduces to an m-dimensional problem, see Theorem 4.1.1 below. In the remaining, we do not discuss the situation $\max \sigma^+(\lambda) = 0$, where the stability behaviour of the reference solution ϕ^* of (D_{λ^*}) is determined by the restriction of (D_λ) on a centre manifold and particularly on Taylor coefficients of nonlinear terms [195]. As opposed to this setting, in the following, stability and bifurcation results are determined by the linear part alone.

The next theorem taken from [189, Theorem 4.14] is a continuous-time and non-linear version of the shovel bifurcation observed in Example 1.2.4.

Theorem 3.2.5 (Shovel Bifurcation) *We suppose that Hypothesis 3.2.4 holds and that the dominant spectral interval $\sigma^-(\lambda)$ has constant multiplicity m^-. Then there exists a neighbourhood $\Lambda_1 \subseteq \Lambda$ of λ^* such that for all $\lambda \in \Lambda_1$, the following statements hold:*

(a) Subcritical case. *If $\lambda \mapsto \max \sigma(\lambda)$ is strictly decreasing at λ^*, then*

- *For $\lambda < \lambda^*$, one has:*
- ○ *If either $\max \sigma^+(\lambda^*) < 0$ or $\lambda \mapsto \max \sigma^+(\lambda)$ strictly increases with $\max \sigma^+(\lambda^*) = 0$, then ϕ_λ is asymptotically stable. If also $\lambda \mapsto \min \sigma^-(\lambda)$ strictly decreases with $\min \sigma^-(\lambda^*) = 0$, then ϕ_λ is embedded into an m^--dimensional family of bounded entire solutions to (D_λ).*
- ○ *If $\lambda \mapsto \min \sigma^+(\lambda)$ strictly decreases with $\min \sigma^+(\lambda^*) = 0$, then ϕ_λ is unstable.*
- *For $\lambda = \lambda^*$ and $\max \sigma^+(\lambda^*) < 0$, the solution ϕ^* is uniformly asymptotically stable on $\mathbb{R}_0^+$.*
- *For $\lambda > \lambda^*$, the unique entire bounded solution of (D_λ) near ϕ_λ is ϕ_λ, and this solution is uniformly asymptotically stable on $\mathbb{R}$.*

(b) Supercritical case. *If $\lambda \mapsto \max \sigma(\lambda)$ is strictly increasing at λ^*, then*

- *For $\lambda < \lambda^*$, the unique entire bounded solution of (D_λ) near ϕ_λ is ϕ_λ, and this solution is uniformly asymptotically stable on $\mathbb{R}$.*
- *For $\lambda = \lambda^*$ and $\max \sigma^+(\lambda^*) < 0$, the solution ϕ^* is uniformly asymptotically stable on $\mathbb{R}_0^+$.*
- *For $\lambda > \lambda^*$, one has:*
- ○ *If either $\max \sigma^+(\lambda^*) < 0$ or $\lambda \mapsto \max \sigma^+(\lambda)$ strictly decreases with $\max \sigma^+(\lambda^*) = 0$, then ϕ_λ is asymptotically stable. If also $\lambda \mapsto \min \sigma^-(\lambda)$ strictly increases with $\min \sigma^-(\lambda^*) = 0$, then ϕ_λ is embedded into an m^--dimensional family of bounded entire solutions to (D_λ).*
- ○ *If $\lambda \mapsto \min \sigma^+(\lambda)$ strictly increases with $\min \sigma^+(\lambda^*) = 0$, then ϕ_λ is unstable.*

The dominant spectral intervals $\sigma(\cdot)$ as functions of λ are illustrated in Fig. 3.4. Note that in the autonomous (or periodic) situation, the classical (or Floquet) spectrum consists of eigenvalues depending differentiably on the parameters [128, Chapter 7]. Since the dichotomy spectrum depends only upper semi-continuously on λ (see Example 5.1.7 below for an example in discrete time), one cannot expect a similar

Figure 3.4 Dominant spectral intervals $\sigma(\lambda)$ as required in Theorem 3.2.5 such that (a) $\max \sigma(\lambda^*) = 0$ is strictly decreasing or (b) $\max \sigma(\lambda^*) = 0$ is strictly increasing

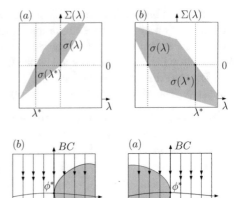

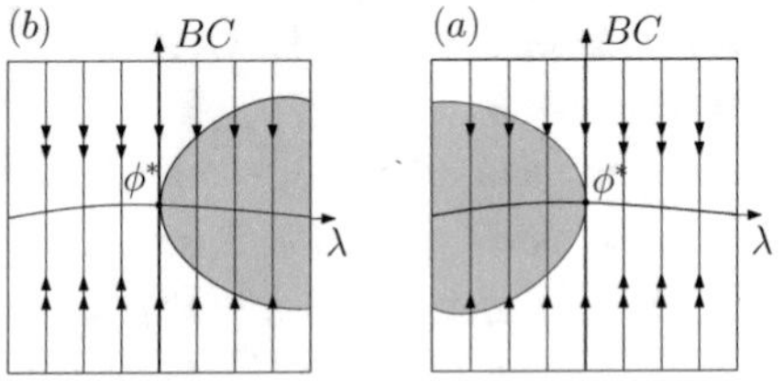

Figure 3.5 Schematic bifurcation diagram for Theorem 3.2.5 yielding a (a) subcritical shovel bifurcation or a (b) supercritical shovel bifurcation of an entire solution ϕ^* (double arrows indicate uniform asymptotic stability on $\mathbb{R}$)

smooth behaviour for the boundary points of spectral intervals. Instead we have to assume monotonicity properties for them.

We refer to Fig. 3.5 for a schematic illustration of the bifurcation patterns described in Theorem 3.2.5. To explain why this phenomenon is called *shovel bifurcation*, we note that the set of bounded solutions for different values of the parameter λ resembles a shovel rather than a pitchfork.

Example 3.2.6 *Consider the asymptotically autonomous equation*

$$\dot{x} = (\lambda + a(t)) \arctan(x) \tag{3.5}$$

with a continuous function $a : \mathbb{R} \to \mathbb{R}$ satisfying

$$\lim_{t \to \infty} a(t) =: \alpha^+ < \alpha^- := \lim_{t \to -\infty} a(t).$$

Along the trivial family $\phi_\lambda(t) \equiv 0$, we refer to Sect. 2.1.1 in order to obtain

$$\Sigma(\lambda) = \left[\lambda + \alpha^+, \lambda + \alpha^-\right], \quad \Sigma^+(\lambda) = \left\{\lambda + \alpha^+\right\}, \quad \Sigma^-(\lambda) = \left\{\lambda + \alpha^-\right\}$$

as the dichotomy spectra from Hypothesis 3.2.4. Their boundary points are strictly increasing in λ. This gives rise to two critical parameter values. At $\lambda^* = -\alpha^-$, the spectral interval enters the stability boundary. This case is covered by Theorem 3.2.5(b) and the trivial solution experiences a transition from uniform asymptotic stability on $\mathbb{R}$ to uniform asymptotic stability on $\mathbb{R}_0^+$ (see Fig. 3.6).

At $\lambda^* = -\alpha^+$, the spectral interval leaves the stability boundary, a situation described in [189, Theorem 4.15]. One observes a transition of the zero solution from uniform asymptotic stability on $\mathbb{R}_0^+$, over stability at $\lambda = -\alpha^-$ to instability (see Fig. 3.7). For $\lambda \in [-\alpha^-, -\alpha^+]$, there exists a family of bounded entire solutions.

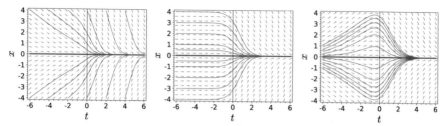

Figure 3.6 Supercritical shovel bifurcation as in Theorem 3.2.5(b): solution curves (blue) of equation (3.5) with coefficient $a(t) = -\tanh t$ for $\lambda = -\frac{3}{2}$ (left, 0 is an isolated bounded entire solution), $\lambda = -1$ (centre) and $\lambda = -\frac{1}{2}$ (right, 0 is embedded in a family of bounded solutions)

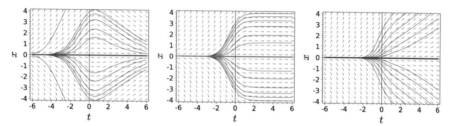

Figure 3.7 Subcritical shovel bifurcation as in Theorem 3.2.5(b): solution curves (blue) of equation (3.5) with coefficient $a(t) = -\tanh t$ for $\lambda = \frac{1}{2}$ (left, 0 is embedded in a family of bounded entire solution), $\lambda = 1$ (centre) and $\lambda = \frac{3}{2}$ (right, 0 is an isolated bounded entire solution)

In the critical situation $\lambda = \lambda^*$, we always enforce asymptotic stability of ϕ^* by the assumption $\max \sigma^+(\lambda^*) < 0$. For the complementary case $\max \sigma^+(\lambda^*) = 0$, one can proceed as follows: reduce the system (D_λ) to a centre manifold (cf. the upcoming Sect. 4.1) and, provided the resulting reduced equation is scalar and ambient assumptions hold, apply a version of Theorem 3.1.4 or 3.1.6. This combines shovel bifurcations with attractor bifurcation phenomena. A concrete example for this interplay is given in [136, Example 3.8].

3.2.3 Rate-Induced Tipping

Rate-induced tipping is a phenomenon where a nonautonomous change of a parameter in an autonomous system leads to a bifurcation of the nonautonomous system when the rate of change of the parameter passes a certain threshold. The simple one-dimensional Example 1.2.5 illustrates this phenomenon clearly, since certain solutions converge to ∞ for rates bigger than a critical rate, while we have convergence to $-\infty$ otherwise.

In this subsection, we describe an elegant formalism to study rate-induced tipping by means of asymptotically autonomous systems. This was first proposed by P. Ashwin, C. Perryman and S. Wieczorek [12] in dimension one and then generalised in

[2, 3] to a higher-dimensional setting. Note that the motivating Example 1.2.5 is not asymptotically autonomous.

We consider an autonomous differential equation

$$\dot{x} = f(x, \lambda), \qquad\qquad (C_\lambda)$$

where $f : \mathbb{R}^d \times [\lambda_-, \lambda_+] \to \mathbb{R}^d$ is continuously differentiable. We assume that there exists a branch of uniformly exponentially stable attractors $A : [\lambda_-, \lambda_+] \to \mathfrak{K}(\mathbb{R}^d)$, i.e. the set $A(\lambda)$ is an invariant set for (C_λ) for all $\lambda \in [\lambda_-, \lambda_+]$, and there exist $\eta, \mu > 0$ and $M \geq 1$ such that

$$h(\phi_\lambda(t, x), A(\lambda)) \leq Ce^{-\mu t} \operatorname{dist}(x, A(\lambda)) \quad \text{for all } x \in B_\eta(M), \, \lambda \in [\lambda_-, \lambda_+],$$

where ϕ_λ denotes the flow of (C_λ). A *parameter shift* from λ_- to λ_+ is a differentiable function $\Gamma : \mathbb{R} \to [\lambda_-, \lambda_+]$ satisfying

  (i) $\lim_{t \to -\infty} \Gamma(t) = \lambda_-, \lim_{t \to \infty} \Gamma(t) = \lambda_+$.
  (ii) $\lim_{t \to \pm\infty} \frac{d\Gamma}{dt}(t) = 0$.

Using this parameter shift and a so-called *rate* $r > 0$, the family of autonomous equations (C_λ) yields a nonautonomous differential equation

$$\dot{x} = f(x, \Gamma(rt)). \qquad\qquad (3.6)$$

This differential equation is asymptotically autonomous, since its right hand side converges to the right hand sides of (C_λ) for $\lambda = \lambda_\pm$ in the limit $t \to \pm\infty$. The following result from [3, Theorem II.2] shows existence of a local pullback attractor that corresponds (in the limit $t \to -\infty$) to the branch A of attractors.

Theorem 3.2.7 (Existence of Local Pullback Attractor) *For all $r > 0$, the differential equation (3.6) has a local pullback attractor $\mathcal{A}_r$ in the sense of (3.1), and the upper backward limit*

$$\overline{\mathcal{A}_r}(-\infty) := \limsup_{t \to -\infty} \mathcal{A}_r(t) := \bigcap_{\tau \leq 0} \overline{\bigcup_{t \leq \tau} \mathcal{A}_r(t)}$$

is contained in $A(\lambda_-)$.

If the rate r is small, then it follows that the branch A is tracked by a local pullback attractor $\mathcal{A}_r$ of (3.6), see [3, Theorem III.1].

Theorem 3.2.8 (Tracking for Small Rates) *For all $\varepsilon > 0$, there exists a $\delta > 0$ such that*

$$h\big(\mathcal{A}_r(t), A(\Gamma(rt))\big) < \varepsilon \quad \text{for all } t \in \mathbb{R} \text{ and } r \in (0, \delta).$$

However, if the rate r is sufficiently large, such a tracking may no longer be possible. In this case, the forward limit of the local pullback attractor will not converge to $A(\lambda_+)$. Here, the *upper forward limit* of the local pullback attractor $\mathcal{A}_r$ is defined by

$$\overline{\mathcal{A}_r}(\infty) := \bigcap_{\tau \geq 0} \overline{\bigcup_{t \geq \tau} \mathcal{A}_r(t)},$$

provided the pullback attractor exists for all $t \in \mathbb{R}$ (in general, its existence is only guaranteed on a half line $(-\infty, a)$).

The following one-dimensional example illustrates rate-induced tipping in the sense that tracking is not possible above a critical rate, and the upper forward limit changes discontinuously. It is essentially an asymptotically autonomous version of Eq. (1.15) from Example 1.2.5 and has been studied in [204, Section II].

Example 3.2.9 *We consider the autonomous differential equation (C_λ) with the right hand side $f(x, \lambda) := (x + \lambda)^2 - 1$ for $\lambda \in [0, 3]$. This differential equation has a branch $A(\lambda) = -1 - \lambda$ of exponentially stable attractors (as well as a branch $R(\lambda) = 1 - \lambda$ of repellers). We consider the parameter shift $\Gamma : \mathbb{R} \to [0, 3]$, given by $\Gamma(t) := \frac{3}{2} \left(\tanh(\frac{3}{2}t) + 1 \right)$. The nonautonomous differential equation (3.6), depending on the rate $r > 0$, then reads as*

$$\dot{x} = (x + \Gamma(rt))^2 - 1 \,.$$

Due to Theorem 3.2.7, for each $r > 0$, there exists a local pullback attractor $\mathcal{A}_r$, which corresponds to the stable branch A and for which we have

$$\lim_{t \to -\infty} \mathcal{A}_r(t) = A(0) = -1 \,.$$

For small $r > 0$, Theorem 3.2.8 guarantees that this pullback attractor is close to the branch A, and in particular, we have for those r that

$$\lim_{t \to \infty} \mathcal{A}_r(t) = A(3) = -4 \,,$$

which implies $\overline{\mathcal{A}_r}(\infty) = \{-4\}$. Note that there exists a corresponding object related to the repelling branch R in terms of a local pullback repeller $\mathcal{R}_r$ (this follows via time reversal from Theorem 3.2.7), and for small $r > 0$, we have

$$\lim_{t \to -\infty} \mathcal{R}_r(t) = R(0) = 1 \quad and \quad \lim_{t \to \infty} \mathcal{R}_r(t) = R(3) = -2 \,.$$

Both attractor and repeller are visualised in Fig. 3.8 (left). One can show that there exists a critical rate $r_c > 0$ for which both attractor and repeller coincide, which implies that we have $\overline{\mathcal{A}_{r_c}}(\infty) = \{-2\}$, see Fig. 3.8 (centre). This means that the set $\overline{\mathcal{A}_r}(\infty)$ changes discontinuously, since $\overline{\mathcal{A}_r}(\infty) = \{-4\}$ for all rates $r \in (0, r_c)$. Note that for all $r > r_c$, the local pullback attractor converges to ∞ in finite time, so $\mathcal{A}_r(\infty)$ is not defined, see Fig. 3.8 (right).

In the literature [3, Definition III.1], several cases of tracking and tipping are discussed.

Definition 3.2.10 (Tracking and Tipping) *Consider* (3.6) *with fixed parameter shift Γ and rate $r > 0$. Let $A : [\lambda_-, \lambda_+] \to \mathfrak{K}(\mathbb{R}^d)$ be a branch of uniformly ex-*

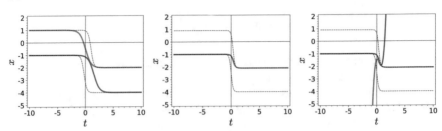

Figure 3.8 Local pullback attractor (in red) and local pullback repeller (in green) of the differential equation discussed in Example 3.2.9; left: tracking for small $r > 0$; centre: at the critical rate, the pullback attractor coincides with the pullback repeller; right: the pullback attractor converges to ∞ in finite time

ponentially stable attractors and $\mathcal{A}_r$ be the corresponding local pullback attractor with upper forward limit $\overline{\mathcal{A}_r}(\infty)$.

    (i) *We say that there is* tracking *for the branch A if $\overline{\mathcal{A}_r}(\infty) \subseteq A(\lambda_+)$, and we distinguish between*

        (a) *Weak tracking, given by $\overline{\mathcal{A}_r}(\infty) \subsetneqq A(\lambda_+)$*

        (b) *Strong tracking, given by $\overline{\mathcal{A}_r}(\infty) = A(\lambda_+)$*

    (ii) *We say that there is* tipping *for the branch A if $\overline{\mathcal{A}_r}(\infty) \not\subseteq A(\lambda_+)$, and we distinguish between*

        (a) *Partial tipping, given by $\overline{\mathcal{A}_r}(\infty) \cap (\mathbb{R}^d \setminus A(\lambda_+)) \neq \emptyset$ and $\overline{\mathcal{A}_r}(\infty) \cap A(\lambda_+) \neq \emptyset$ and*

        (b) *Total tipping, given by $\overline{\mathcal{A}_r}(\infty) \cap A(\lambda_+) = \emptyset$.*

We note that in Example 3.2.9, there is strong tracking for the branch A for $r < r_c$ and total tipping for $r \geq r_c$. For examples of weak tracking and partial tipping, we refer the reader to [2, 3].

3.3 Bifurcation of Minimal Sets

In this section, we discuss several nonautonomous counterparts to the classical autonomous bifurcations of fold, transcritical and Hopf type in the context of skew product flows. This allows to compare the bifurcation results obtained from the process and skew product formulation. In contrast to the previous scenarios, now the bifurcating objects (here, minimal sets) reflect the dynamics of the driving system in the resulting skew product flow. This means that a skew product formulation can give further insights into the bifurcating objects. It yields more information compared to for example the rather crude shovel bifurcation, or the pure existence of bounded entire solutions or nontrivial attractors/repellers. This benefit requires a certain familiarity with notions from ergodic theory [129, 218].

3.3.1 Scalar Equations

We begin with our standing assumptions on the driving dynamics. Suppose that a dynamical system $\theta : \mathbb{R} \times \Omega \to \Omega$ on a compact metric space Ω is strictly ergodic (with respect to a probability space $(\Omega, \mathfrak{B}, \mu)$).

First, we describe an approach to transcritical bifurcations due to R.A. Johnson and F. Mantellini [122], which is based on averaging techniques. Its initial point is an autonomous triangular system

$$\begin{cases} \dot{y} = g(y), \\ \dot{x} = |\lambda| \, f_0(y, x, \lambda) \end{cases} \tag{3.7}$$

depending on a parameter $\lambda \in \Lambda$ from a compact neighbourhood $\Lambda \subset \mathbb{R}$ of 0. The mapping g is assumed to be sufficiently smooth to generate a minimal and uniquely ergodic flow $\theta : \mathbb{R} \times \Omega \to \Omega$ on a compact subset $\Omega \subset \mathbb{R}^m$. For the right hand side $f_0 : \Omega \times \mathbb{R} \times \Lambda \to \mathbb{R}$ in (3.7), we assume that the partial derivatives $D_2^j f_0$ exist as continuous functions for $0 \leq j \leq 3$. Moreover, we suppose that

$$f_0(\omega, 0, \lambda) = 0 \quad \text{for all } \omega \in \Omega, \, \lambda \in \Lambda. \tag{3.8}$$

This gives rise to driven scalar nonautonomous differential equations

$$\dot{x} = f(\theta(t, \omega), x, \lambda), \tag{D_λ^ω}$$

whose solution satisfying the initial condition $x(0) = \xi$ is denoted as $\varphi_\lambda(\cdot, \omega, \xi)$.

The present approach [122] requires right hand sides of the form

$$f(\omega, x, \lambda) := |\lambda| \, f_0(\omega, x, \lambda). \tag{3.9}$$

The set-up is designed so that the trivial solution to (D_λ^ω) loses asymptotic stability when λ passes from negative to positive values at least for some $\omega \in \Omega$. It is natural to ask for alternative invariant objects to which the stability is transferred to.

Thanks to the unique ergodicity of θ, one obtains from Birkhoff's ergodic theorem [169, pp. 459ff] that the mean values

$$\bar{a}(\lambda) := \int_\Omega D_2 f_0(\omega, 0, \lambda) \, d\mu(\omega) = \lim_{t \to \infty} \frac{1}{t} \int_0^t D_2 f_0(\theta(s, \omega), 0, \lambda) \, ds,$$

$$\bar{b}(\lambda) := \int_\Omega D_2^2 f_0(\omega, 0, \lambda) \, d\mu(\omega) = \lim_{t \to \infty} \frac{1}{t} \int_0^t D_2^2 f_0(\theta(s, \omega), 0, \lambda) \, ds$$

are well defined, where the limits are uniform in $\omega \in \Omega$, $\lambda \in \Lambda$.

The subsequent result summarises [122, Theorems 2.1 and 2.2] and requires the notion of a *local attractor* $\mathcal{A}_\lambda$ in the present skew product setting. This means that there exists an open neighbourhood $U \subseteq \mathbb{R}$ in the sense of $\mathcal{A}_\lambda \subset \Omega \times U$ satisfying

$$\Omega \times \bigcap_{T>0} \bigcup_{t \geq T} \varphi_\lambda(t, \Omega, U) = \mathcal{A}_\lambda.$$

Theorem 3.3.1 (Transcritical Bifurcation) *Suppose that* (3.8) *and the nonhyperbolicity condition* $\bar{a}(0) = 0$ *and* $\bar{b}(0) < 0$ *are satisfied. For every* $R > 0$, *there exist reals* $\lambda_0 > 0$ *in* Λ *and* $r > 0$ *such that the solution* $\varphi_\lambda(\cdot, \omega, \xi)$ *exists on* $\mathbb{R}_0^+$ *and satisfies* $\varphi_\lambda(t, \omega, \xi) \in \bar{B}_R(0)$ *for all* $t \geq 0$, $\lambda \in (0, \lambda_0]$, $\omega \in \Omega$ *and* $\xi \in \bar{B}_r(0)$. *In addition, the following properties hold for all* $\lambda \in (0, \lambda_0]$:

(a) *For each* $\omega \in \Omega$ *and* $\xi \in (0, r]$, *the forward limit set* $L^+(\omega, \xi) \subset \Omega \times \mathbb{R}$ *of* (3.7) *contains at least one and at most two minimal subsets.*

(b) *For those* $\omega \in \Omega$ *fulfilling the inequality*

$$\left(1 + \sup_{\substack{\omega \in \Omega, \\ \lambda \in \Lambda}} |D_2 f_0(\omega, 0, \lambda)|\right) \sup_{\omega \in \Omega} \left| \lambda \int_0^\infty e^{-\lambda s} \left[D_2 f_0(\theta(s, \omega), 0, \lambda)\right] - \bar{a}(\lambda)\right| \, ds$$

$$\leq \tfrac{1}{10}\bar{a}(\lambda), \tag{3.10}$$

the system (3.7) *has a local attractor* $\mathcal{A}_\lambda \subseteq \Omega \times \mathbb{R}$ *disjoint from* $\Omega \times \{0\}$. *Moreover, to each* $\varepsilon > 0$ *corresponds a* $\lambda_\varepsilon \in (0, \lambda_0]$ *such that if* (3.10) *holds at* $\lambda \in (0, \lambda_\varepsilon]$, *then* $d(\mathcal{A}_\lambda, \Omega \times \{0\}) < \varepsilon$.

As indicated by [122, Lemma 2.1], the left hand side in (3.10) tends to 0 for $\lambda \searrow 0$. The minimal subsets guaranteed in statement (i) are *almost automorphic extensions* of the skew product flow $(\theta, \varphi_\lambda)$ (see [122]). It ensures that (3.7) possesses at least one minimal set in $\Omega \times \mathbb{R}$. This set is possibly given by $\Omega \times \{0\}$, which can happen in case $\bar{a}(\lambda) > 0$ for $\lambda > 0$. Under the additional condition (3.10), however, there exists a local attractor $\mathcal{A}_\lambda$ as 'new' compact invariant set different from the trivial one $\Omega \times \{0\}$.

We note that, due to the multiplicative parameter dependence in (3.9), the above Theorem 3.3.1 does not apply to our previous Examples 3.1.5 or 3.2.3 for nonautonomous transcritical bifurcations. Sufficient conditions for also fold bifurcations in (D_λ^ω) in a similar setting are given in [77, 78]. Finally, fold, transcritical and pitchfork patterns where the bifurcation parameter is subject to fast oscillations are discussed in [82].

Concerning nonautonomous fold bifurcations, we next follow an alternative approach from C. Núñez and R. Obaya [173]. It allows a driving flow $\theta : \mathbb{R} \times \Omega \to \Omega$ on a general compact metric space Ω. It drives a scalar nonautonomous differential equation (D_λ^ω). The right hand side $f : \Omega \times \mathbb{R} \times \mathbb{R} \to \mathbb{R}$ exists as a continuous function together with the partial derivative $D_2 f$ and satisfies global assumptions.

Hypothesis 3.3.2 *We suppose the following:*

(i) $f(\omega, \cdot, \lambda)$ *is concave for* $(\omega, \lambda) \in \Omega \times \mathbb{R}$, *and for all* $\lambda \in \mathbb{R}$, *there exists an* $\Omega_\lambda \subseteq \Omega$ *with* $\mu(\Omega_\lambda) > 0$ *so that* $f(\omega, \cdot, \lambda)$ *is strictly concave for* $\omega \in \Omega_\lambda$.

(ii) $f(\omega, x, \cdot)$ *is strictly increasing for all* $(\omega, x) \in \Omega \times \mathbb{R}$, *and there exist real numbers* $\delta, r_1 > 0$ *and* $\lambda_+, \lambda_- \in \mathbb{R}$ *such that*

$$f(\omega, r_1, \lambda_+) > 0, \qquad f(\omega, x, \lambda_-) < -\delta \quad \text{for all } \omega \in \Omega \text{ and } x \in \mathbb{R}.$$

(iii) $\lim_{x\to\infty} f(\omega, x, \lambda) = -\infty$ *for all* $(\omega, \lambda) \in \Omega \times \mathbb{R}$.

For instance, the above assumptions hold for quadratic functions $f(\omega, x, \lambda) := a(\omega, \lambda)x^2 + b(\omega, \lambda)x + c(\omega, \lambda)$ with continuous coefficients $a, b, c : \Omega \times \mathbb{R} \to \mathbb{R}$, whose partial derivatives with respect to the second variable exist as continuous functions and which satisfy

$$a(\omega, \lambda) < 0, \qquad\qquad D_2 a(\omega, \lambda) > 0,$$
$$4D_2 a(\omega, \lambda) D_2 c(\omega, \lambda) > D_2 b(\omega, \lambda)^2,$$
$$c(\omega, \lambda_+) > \frac{b(\omega,\lambda_+)^2}{4a(\omega,\lambda_+)}, \qquad\qquad c(\omega, \lambda_-) < \frac{b(\omega,\lambda_-)^2}{4a(\omega,\lambda_-)}$$

for all $\omega \in \Omega, \lambda \in \mathbb{R}$.

The next result due to [173, Section 3] indicates a supercritical fold bifurcation. For $\lambda < \lambda^*$, there exists no minimal set, at the critical parameter $\lambda = \lambda^*$ there is a minimal nonhyperbolic set and finally, for $\lambda > \lambda^*$, there are two minimal sets.

Theorem 3.3.3 (Supercritical Fold Bifurcation of Minimal Sets) *Suppose that Hypothesis 3.3.2 holds. Then for all $\lambda \in \mathbb{R}$, the nonautonomous set*

$$\mathcal{B}_\lambda := \{(\omega, \xi) \in \Omega \times \mathbb{R} : \varphi_\lambda(\cdot, \omega, \xi) \text{ is a bounded entire solution of } (D_\lambda^\omega)\}$$

is compact and $(\theta, \varphi_\lambda)$-invariant. Moreover, there exists a critical parameter value $\lambda^* \in \mathbb{R}$ such that the following is true for the functions $\phi_\lambda^\pm : \Omega \to \mathbb{R}$:

$$\phi_\lambda^+(\omega) := \sup\{x : (\omega, x) \in \mathcal{B}_\lambda\}, \qquad \phi_\lambda^-(\omega) := \inf\{x : (\omega, x) \in \mathcal{B}_\lambda\}$$

defining $(\theta, \varphi_\lambda)$-invariant curves:

(a) If $\lambda < \lambda^*$, then $\mathcal{B}_\lambda = \emptyset$, and no subsets of $\Omega \times \mathbb{R}$ are $(\theta, \varphi_\lambda)$-minimal.

(b) If $\lambda = \lambda^*$, then $\mathcal{B}_{\lambda^*} \neq \emptyset$, and

$$\lim_{\lambda \searrow \lambda^*} \phi_\lambda^\pm(\omega) = \phi_{\lambda^*}^\pm(\omega) \quad \text{for all } \omega \in \Omega.$$

The semi-continuous functions $\phi_{\lambda^*}^\pm$ coincide in a residual set $R_{\lambda^*} \subseteq \Omega$, which is θ-invariant and $\mathcal{B}_{\lambda^*}$ is a pinched set, that is, for every $\omega \in R_{\lambda^*}$ the fibre $\mathcal{B}_{\lambda^*}(\omega)$ is a singleton. The skew product flow $(\theta, \varphi_{\lambda^*})$ has a minimal, nonhyperbolic set ϕ^*, and the following scenarios can occur:

- *$R_{\lambda^*} = \Omega$, which happens when $\phi_\lambda^\pm$ collapse in the entire base as $\lambda \searrow \lambda^*$ and in which case $\phi^* = \mathcal{B}_{\lambda^*}$ is a copy of the base.*
- *$R_{\lambda^*} \neq \Omega$ and $\mu(R_{\lambda^*}) = 1$, in which case $\mathcal{B}_{\lambda^*}$ is not a copy of the base.*
- *$R_{\lambda^*} \neq \Omega$ and $\mu(R_{\lambda^*}) = 0$, in which case neither $\mathcal{B}_{\lambda^*}$ nor $\phi^* \subseteq \mathcal{B}_{\lambda^*}$ is a copy of the base.*

(c) If $\lambda > \lambda^*$, then $\mathcal{B}_\lambda \neq \emptyset$. The nonautonomous sets $\phi_\lambda^\pm$ are the only sets being $(\theta, \varphi_\lambda)$-minimal. They are hyperbolic in terms of Lyapunov exponents

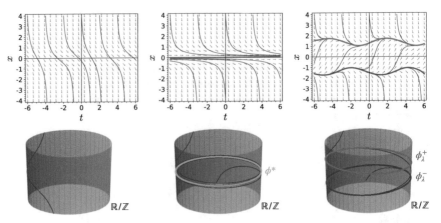

Figure 3.9 Supercritical fold bifurcation as in Theorem 3.3.3 in the spaces $\mathbb{R} \times \mathbb{R}$ (top) and $(\mathbb{R}/\mathbb{Z}) \times \mathbb{R}$ (bottom): solution curves (blue) of the equation (3.11) for (**a**) $\lambda = -1$ (left, $\mathcal{B}_\lambda = \emptyset$), (**b**) $\lambda = 0$ (centre, $\mathcal{B}_0 = \mathbb{R}/\mathbb{Z} \times \{0\}$) and (**c**) $\lambda = 2$ (right, $\mathcal{B}_\lambda$ consists of the solutions between the attractive periodic solutions ϕ_λ^+ (green) and the repelling periodic solution ϕ_λ^- (red))

$\chi_\mu(\phi_\lambda^+) < 0 < \chi_\mu(\phi_\lambda^-)$ *and vary continuously in* λ*. In particular, the functions* $(\omega, \lambda) \mapsto \phi_\lambda^\pm(\omega)$ *are continuous on* $\Omega \times (\lambda^*, \infty)$ *and* $\phi_\lambda^-(\omega)$ *and* $-\phi_\lambda^+(\omega)$ *are strictly decreasing in* $\lambda > \lambda^*$*.*

We illustrate Theorem 3.3.3 using a rather simple example on the n-torus:

Example 3.3.4 (Quasi-Periodic Riccati Equation) *Let* θ *denote the Kronecker flow on the* n*-torus* $\Omega = \mathbb{R}^n/\mathbb{Z}^n$ *from Example 1.1.1 and suppose that* $g : \Omega \to (0, \infty)$ *is a continuous function. We consider the nonautonomous Riccati equation*

$$\dot{x} = \lambda g(\theta(t, \omega)) - x^2,$$

which is of the form (D_λ^ω) *with* $f(\omega, x, \lambda) := \lambda g(\omega) - x^2$*. Then Hypothesis 3.3.2 holds with* $\Omega_\lambda = \mathbb{R}^n/\mathbb{Z}^n$*, e.g.* $\delta := \inf_{\omega \in \Omega} g(\omega) > 0$*,* $r_1 := 0$*,* $\lambda_- := -1$*,* $\lambda_+ := 1$ *and thus Theorem 3.3.3 applies with the critical parameter* $\lambda^* = 0$*. Indeed, for* $\lambda < 0$*, all solutions are unbounded* $(\mathcal{B}_\lambda = \emptyset)$*. In the nonhyperbolic case* $\lambda = 0$*, one has* $\mathcal{B}_0 = \Omega \times \{0\} = \phi^*$*, which is a copy of the base. Finally, for parameters* $\lambda > 0$*, the two sets* $\phi_\lambda^-, \phi_\lambda^+$ *arise, both represent quasi-periodic solutions and the remaining set* $\mathcal{B}_\lambda$ *consists of solutions in between connecting the repelling* ϕ_λ^- *with the attracting* ϕ_λ^+*. In order to visualise this scenario, we retreat to the situation* $\Omega = \mathbb{R}/\mathbb{Z}$ *and* $g(\omega) := 1 + \frac{\operatorname{Im} \omega}{2}$*,* $\theta(t, \omega) := \omega + e^{it}$*. The behaviour of the resulting* 2π*-periodic differential equation*

$$\dot{x} = \lambda(1 + \tfrac{1}{2} \sin t) - x^2 \tag{3.11}$$

is illustrated in Fig. 3.9.

Finally, the reference [173] also contains criteria for nonautonomous transcritical and pitchfork bifurcations in the above framework.

3.3.2 Hopf Bifurcation

In this chapter on continuous-time nonautonomous bifurcations we focus exclusively on the one-dimensional case. Bifurcations in higher-dimensional systems are covered in Chap. 4 by using reduction principles. Moreover, we look at a nonautonomous version of the Sacker-Neimark bifurcation in the discrete-time setting of Chap. 6.

We note that a prototypical, yet characteristic, example for a nonautonomous Hopf bifurcation was already given in Example 1.2.7, and there are a few further results on Hopf bifurcations in the literature, which we do not cover in detail due to the technical complexity of these results. These results date back to the early 1990s, when R.A. Johnson and Y. Yi [120, 124] investigated Hopf bifurcations from non-periodic solutions of differential equations. Recent investigations address bifurcation parameters subject to fast oscillations [82] or include a detailed description of the topological dynamics of a nonautonomous differential equations driven by a Kronecker flow [83] and the verification of a nonautonomous version of Li–Yorke chaos at a Hopf bifurcation point, i.e. in a nonhyperbolic situation [174].

Remarks

The paper [140] by P.E. Kloeden and S. Siegmund and also parts of the monographs [43, 139] serve as introduction to several aspects of nonautonomous bifurcations. Moreover, we point out that except Theorem 3.3.3 (and the entire approach in [173]), all our bifurcation results are of local nature.

Attractor and Solution Bifurcation Autonomous transcritical and pitchfork bifurcations fit into the framework of Theorems 3.1.4 and 3.1.6 (for this, see [200, Example 5.3, respectively, 6.3]. However, it seems that an autonomous fold bifurcation is not suitable for a formulation in terms of an attractor bifurcation.

The monograph [163, pp. 114ff, Section 5.2] suggests a concept of attractor bifurcations for autonomous differential equations and a generalisation to nonautonomous problems would be interesting. The contribution [135] might help to establish a connection between solution and attractor bifurcation. It is shown that every compact forward (or backward) invariant set contains a strictly invariant nonautonomous set. Since attractors consist of bounded entire solutions [43, p. 15, Corollary 1.18], attractor bifurcation will lead to solution bifurcation.

Information on the number of bounded entire solutions to scalar differential equations was given in [29, 30], with immediate implications for solution bifurcations.

Persistence results describing the behaviour of bifurcation diagrams for solution bifurcations under perturbation are given in [191].

Rate-induced tipping with nonautonomous limiting systems has been studied in [160, 161].

Shovel Bifurcation Apparently the classical autonomous transcritical and pitchfork patterns can be interpreted as 'shovel bifurcations' in the sense that a whole family of bounded entire solutions branches off, namely the heteroclinic connections of the bifurcating fixed points. However, the corresponding assumptions significantly differ from Hypothesis 3.2.4. Since the attractor bifurcation Theorems 3.1.4 and 3.1.6 apply to equations on half axes $\mathbb{R}_0^+$ and $\mathbb{R}_0^-$ [199, 200], they can support Theorem 3.2.5 in order to classify the behaviour at the critical parameters. On a linear basis, shovel bifurcations are transitions from exponential dichotomies to exponential trichotomies; the latter notion is due to [73].

Topological Methods Although topological methods are a standard tool in analytical bifurcation (branching) theory [60, pp. 398ff, § 29], [132, pp. 195, Chapter II] and [229, pp. 653ff, Chapter 15], they are rarely applied to nonautonomous equations. One reason is that e.g. the Leray–Schauder degree necessitates too restrictive assumptions in case of aperiodic time-dependencies. Nonetheless, topological arguments proved successful to find bifurcation points for bounded solutions to autonomous differential equations in [113], while [226] applies Conley index theory to almost periodic equations. In particular, the recent nonautonomous Conley index theory [115–117] might be fruitful to locate bifurcation points in time-variant problems.

Quasi- and Almost Periodic Bifurcations Results for the bifurcation of invariant tori [47, 214] already involve tools and ideas from nonautonomous dynamics. There is a well-developed bifurcation theory for quasi-periodic differential equations whose frequencies satisfy Diophantine conditions [38, 39]. The proofs heavily rely on KAM and normal form theory and require smooth coefficients. Without Diophantine conditions, so-called bubbles in the parameter space arise, in which very limited information on the bifurcating objects is at hand. Yet, in such parameter regimes, Theorem 3.3.1 can apply [122] (see also [77]).

A continuous-time version of the subsequent Theorem 6.3.2 (found in [5, Theorem 7.1]) can be seen as a generalisation of results in [172] and [173], detailed in Sect. 3.3, although the methods of proof are quite different. We also refer to the fold bifurcations in quasi-periodically driven differential equations from [87].

Bifurcation of solutions to almost periodic equations is studied in [134, 142] and [226]. We refer to [89] for results on almost periodic variational equations.

Miscellaneous One possibility to ensure nonhyperbolicity $0 \in \Sigma(\lambda^*)$ is to violate the statement (b) from Theorem 2.1.4. This results in an abstract branching theory based on Fredholm properties [132, 229], and we refer to the analogous discrete-time approach in Sects. 6.2.1 and 6.2.2. Sufficient conditions for resulting fold, transcritical and pitchfork bifurcation patterns can be found in [187, Section 3].

A time-variant toy SIR model from epidemics is studied in [136] in order to compare these different approaches [152, 173, 189, 200] to a nonautonomous bifurcation theory. More recently, [203] investigates the influence of external forcing on fold bifurcations and corresponding early warning signals. Concerning bifurcation results in control flows, we refer to [51].

Although we restrict to finite-dimensional differential equations, bifurcation phenomena were also discussed in nonautonomous evolutionary partial differential equations. A logistic differential equation with diffusion and a time-variant coefficient in the nonlinear term is studied in [151]; they establish that a unique bounded entire solution bifurcates from zero at a critical parameter value. Furthermore, [42] addresses nonautonomous pitchfork bifurcations in reaction-diffusion equations.

Chapter 4
Reduction Techniques

Various results and methods in bifurcation theory are based on simplifications of parametrised differential equations. They fall basically into two categories:

- *Dimension reduction.* In the functional analytical approach to nonautonomous bifurcations from Sects. 3.2 and 6.2, the method of Lyapunov–Schmidt allows a reduction of an abstract equation in a function (or sequence) space to a finite-dimensional algebraic equation. The time-dependence in (D_λ) has an impact on these spaces, but not on the method itself. Centre manifolds provide an alternative approach for dimension reduction of bifurcation problems (D_λ) which, in addition, is dynamically meaningful [228, pp. 245ff, Chapter 18]. There exist locally invariant graphs in the state space capturing the essential dynamics of a system. A comparison between these two techniques is given in [132, p. 24, Remark I.7.1].
- *Algebraic simplification.* Normal form theory transforms differential equations to their algebraically most simple or essential form [228, pp. 270ff, Chapter 19]. Provided so-called nonresonance conditions hold, it allows to remove monomial terms not required for a given bifurcation behaviour.

Also symmetries of a dynamical system often allow to identify more specific bifurcation phenomena or algebraic simplifications. However, due to a lack of references or a systematic approach in the nonautonomous theory, we merely mention them.

This chapter tackles the corresponding reduction theories of centre manifolds and normal forms for time-variant differential equations (D). By means of an explicit example, we demonstrate a nonautonomous bifurcation analysis based on centre manifold reduction.

Throughout, let us consider smooth right hand sides $f : \mathcal{D} \to \mathbb{R}^d$ and we are interested in the behaviour near a fixed reference solution $\phi^* : \mathbb{I} \to \mathbb{R}^d$ defined on an unbounded interval $\mathbb{I}$ and satisfying $\mathcal{B}_R(\phi^*) \subseteq \mathcal{D}$ for some $R > 0$. Our analysis is based on the equation of perturbed motion

$$\dot{x} = D_2 f(t, \phi^*(t))x + F(t, x)$$

© The Author(s), under exclusive license to Springer Nature Switzerland AG 2023
V. Anagnostopoulou et al., *Nonautonomous Bifurcation Theory*, Frontiers
in Applied Dynamical Systems: Reviews and Tutorials 10,
https://doi.org/10.1007/978-3-031-29842-4_4

with the nonlinearity $F : \mathbb{I} \times B_R(0) \to \mathbb{R}^d$, given by

$$F(t, x) := f(t, x + \phi^*(t)) - f(t, \phi^*(t)) - D_2 f(t, \phi^*(t))x\,.$$

We assume the limit relation

$$\lim_{x \to 0} D_2 F(t, x) = 0 \quad \text{uniformly in } t \in \mathbb{I}$$

and that the variational equation

$$\dot{x} = D_2 f(t, \phi^*(t))x\,, \tag{V}$$

whose transition matrix is denoted by $\Phi(t, s)$, has the dichotomy spectrum

$$\Sigma_{\mathbb{I}}(\phi^*) = \bigcup_{i=1}^{n} [a_i, b_i] \tag{4.1}$$

with boundary points $b_i < a_{i+1}$ for $1 \le i < n$.

4.1 Centre Integral Manifolds

There are two reasons for the importance of centre manifolds in bifurcation theory. First, by means of the reduction principle in Theorem 4.1.1, they allow to determine stability properties of nonhyperbolic solutions, i.e. in case of critical parameter values. Second, centre manifolds contain solutions staying close to reference solutions, that is bifurcating objects. Therefore, reducing a differential equation to a centre manifold enables us to make use of the lower-dimensional bifurcation results from the previous sections in a general higher-dimensional framework.

We present some basics on the properties and local approximations of centre manifolds in this section. Since the theory of nonautonomous centre (integral) manifolds is presented more comprehensively in [139, pp. 105ff, Chapter 6], only the essential ingredients are provided here.

Let us first consider nonautonomous differential equations (D) with a C^m-right hand side f, $m \ge 2$, and solutions ϕ^* on an interval $\mathbb{I}$ being unbounded below. In order to describe local integral manifolds, we assume there exist real numbers $\beta_+ < \beta_-$ satisfying a *gap condition*

$$\Sigma_{\mathbb{I}}(\phi^*) \cap (\beta_+, \beta_-) = \emptyset, \qquad\qquad \beta_+ < m\beta_-\,.$$

Choosing $\gamma \in (\beta_+, \beta_-)$, the associated scaled equation (L_γ) has an exponential dichotomy, whose invariant projector is called $P : \mathbb{I} \to \mathbb{R}^{d \times d}$. Because $\mathbb{I}$ is unbounded below, the associated unstable set $\mathcal{V}^-$ is uniquely determined due to Proposition 2.1.2.

If $U \subseteq \mathbb{R}^d$ is an open and convex neighbourhood of 0 and $c : \mathbb{I} \times U \to \mathbb{R}^d$ a continuously differentiable function satisfying

$$c(t,0) \equiv 0 \quad \text{on } \mathbb{I}, \qquad \lim_{x \to 0} D_2 c(t,x) = 0 \quad \text{uniformly in } t \in \mathbb{I},$$

$$c(t,x) = c\big(t, (\mathrm{id} - P(t))x\big) \in R(P(t)) \quad \text{for all } t \in \mathbb{I}, \ x \in U,$$

then its graph

$$\phi^* + \mathcal{C} := \big\{ (\tau, \phi^*(\tau) + \xi + c(\tau,\xi)) \in \mathbb{I} \times \mathbb{R}^d : \xi \in N(P(\tau)) \cap U \big\}$$

is called *locally invariant* for the solution ϕ^* to (D), provided (see Fig. 4.1)

$$(t_0, x_0) \in \phi^* + \mathcal{C} \quad \Rightarrow \quad (t, \varphi(t,t_0,x_0)) \in \phi^* + \mathcal{C} \quad \text{for all } t_0 \leq t$$

holds, as long as the general solution φ to (D) satisfies $\varphi(t,t_0,x_0) \in \phi^* + U$.

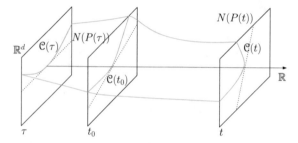

Figure 4.1 Fibres $\mathcal{C}(t)$ of an integral manifold $\mathcal{C} \subseteq \mathcal{D}$ along the trivial solution being smooth curves tangential to the kernels $\mathcal{V}^-(t) = N(P(t)), t \in \mathbb{I} \subseteq \mathbb{R}$

Then [194, Theorem 3.2] guarantees that the solution ϕ^* to (D) possesses a locally invariant integral manifold $\mathcal{C}$ as graph of a function c, whose partial derivatives $D_2^n c$ exist and are continuous for $1 \leq n \leq m$.

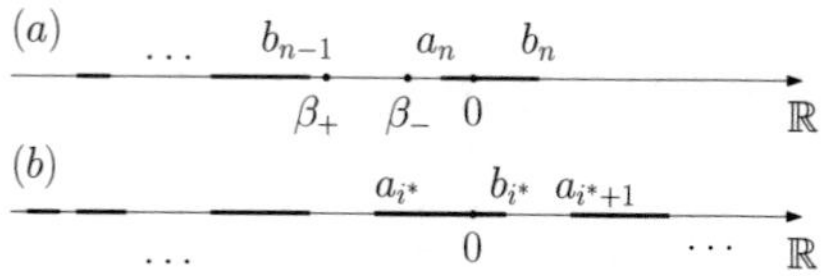

(a)
$$\ldots \quad b_{n-1} \quad a_n \quad b_n$$
$$\beta_+ \quad \beta_- \quad 0 \qquad\qquad \mathbb{R}$$

(b)
$$a_{i^*} \quad b_{i^*} \ a_{i^*+1}$$
$$\ldots \qquad 0 \qquad \ldots \quad \mathbb{R}$$

Figure 4.2 (a) Dichotomy spectrum required in Theorem 4.1.1. (b) More general and unstable situation

From now on, we restrict to the situation of Fig. 4.2a, where the dominant spectral interval $[a_n, b_n]$ contains 0, while β_- chosen according to $b_{n-1} < \beta_- < a_n$. In this case $\mathcal{C}$ defines a *centre (integral) manifold*. Because of attractiveness properties of $\mathcal{C}$, the mentioned dimension reduction via centre manifolds manifests itself in the next result [139, p. 127, Theorem 6.25].

Theorem 4.1.1 (Reduction Principle) *Suppose that* $\mathbb{I} = \mathbb{R}$ *and* $\beta_- < 0$. *A solution* ϕ^* *of* (D) *is stable (uniformly stable, asymptotically stable, uniformly asymptotically stable, exponentially stable, uniformly exponentially stable or unstable) if and only if the trivial solution of the* reduced equation

$$\dot{x} = D_2 f(t, \phi^*(t))x + (\mathrm{id} - P(t))F(t, x + c(t, x)) \qquad (4.2)$$

in $\mathcal{V}^-$ *has the respective stability property.*

We note here that (4.2) is a nonautonomous differential equation in the spectral manifold $\mathcal{V}^- = \mathcal{V}_n$ from the Spectral Theorem 2.1.6. Hence, the dimension of the reduced equation (4.2) is the multiplicity of the dominant spectral interval $[a_n, b_n]$.

In order to analyse the reduced differential equation (4.2), one needs to know the function c at least in form of a Taylor approximation

$$c(t, x) = \sum_{n=2}^{m} \frac{1}{n!} c_n(t) x^{(n)} + R_m(t, x)$$

with coefficients $c_n : \mathbb{I} \to L_n(\mathbb{R}^d)$, $c_n(t) := D_2^n c(t, 0)$ and a remainder R_m satisfying $\lim_{x \to 0} \frac{R_m(t,x)}{\|x\|^m} = 0$ uniformly in $t \in \mathbb{I}$. As shown in [194, Theorem 4.2(b)], the Taylor coefficients are determined recursively from *Lyapunov–Perron integrals*

$$c_n(t) = \int_{-\infty}^{t} \Phi(t, s) H_n(s)|_{\Phi(s,t)(\mathrm{id} - P(t))} \, ds \quad \text{for all } t \in \mathbb{I}, \, 2 \leq n \leq m. \quad (4.3)$$

Understanding this formula requires to define the *partitions* of length two, namely

$$P_2(n) := \left\{ (N_1, N_2) \, \middle| \, \begin{array}{l} N_1, N_2 \subseteq \{1, \ldots, n\} \text{ and } N_1, N_2 \neq \emptyset, \\ N_1 \cup N_2 = \{1, \ldots, l\}, \, N_1 \cap N_2 = \emptyset \end{array} \right\}$$

and the ordered partitions $P_l^<(n)$ introduced in Sect. 2.3. Here, the multilinear mappings $H_n(t) \in L_n(\mathbb{R}^d)$ are given as

$$H_n(t)x_1 \cdots x_n := P(t) D_2^n F(t, 0)x_1 \cdots x_n$$

$$+ (\mathrm{id} - P(t)) \sum_{l=2}^{n-1} \sum_{(N_1, \ldots, N_l) \in P_l^<(n)} D_2^l F(t, 0) C_{\#N_1}(t)x_{N_1} \cdot C_{\#N_l}(t)x_{N_l}$$

$$- \sum_{\substack{(N_1, N_2) \in P_2(n) \\ 0 < \#N_1 < n-1 \\ N_2 \neq \emptyset}} c_{\#N_1 + 1}(t)x_{N_1} \cdots g_{\#N_l}(t)x_{N_2}$$

and particularly $H_2(t) = (\mathrm{id} - P(t)) D_2^2 F(t, x)$. We have used the abbreviations $C(t, x) := \mathrm{id} - P(t))x + c(t, x)$, $g(t, x) := D_2 f(t, \phi^*(t))P(t)x + P(t)F(t, C(t, x))$ and the higher order chain rule from [139, p. 119, Lemma 6.16] implies

$$g_1(t)x_1 = D_2 f(t, \phi^*(t))P(t)x_1$$

and for all $2 \leq n \leq m$ and $x_1, \ldots, x_n \in \mathbb{R}^d$ that

$$g_n(t)x_1 \cdots x_n =$$

$$\sum_{j=2}^{n} \sum_{(N_1,\ldots,N_j)\in P_j^<(n)} P(t)D_2^j F(t,0)C_{\#N_1}(t)_{P(t)}x_{N_1} \cdots C_{\#N_j}(t)_{P(t)}x_{N_j},$$

where $C_n(t) := D_2^n C(t,0)$. Observe that the values $H_n(t)$ only depend on the functions $c_2, \ldots, c_{n-1}$. Yet, in actual computation, we recommend working with the invariance equation [139, p. 118, (6.15)] for $\mathcal{C}$, rather than evaluating $H_2(t)$ etcetera.

When applying the above analysis to parameter-dependent problems (D_λ), one is confronted with the following problem: the centre manifold only exists for parameters, where the dominant spectral interval contains 0. This possibly only holds in case $\lambda = \lambda^*$ when the dominant interval is a singleton or shrinks suddenly. In order to obtain centre manifolds for all λ from an open neighbourhood of the critical value λ^*, one considers the *augmented differential equation*

$$\begin{cases} \dot{x} = f(t,x,\lambda), \\ \dot{\lambda} = 0. \end{cases} \tag{$\bar{D}_\lambda$}$$

Here the parameter λ is interpreted as state space variable in addition to x. Now our above approach yields λ-dependent centre manifolds existing for λ near λ^*. Then the reduced equation (4.2) depends on λ and is called *bifurcation equation*.

The following example illustrates this procedure and how the previous bifurcation results apply to the bifurcation equation.

Example 4.1.2 (Nonautonomous Lorenz Equations) *Let us consider a non-autonomous version of the well-known* Lorenz equations *from, e.g. [148, pp. 188, 291] for illustrative, rather than truly physical reasons. Given by a three-dimensional system*

$$\begin{cases} \dot{x}_1 = \sigma_\lambda(t)(x_2 - x_1), \\ \dot{x}_2 = \rho_\lambda(t)x_1 - x_2 - x_1 x_3, \\ \dot{x}_3 = -\beta_\lambda(t)x_3 + x_1 x_2, \end{cases} \tag{4.4}$$

which obviously can be written in the form (D_λ) with right hand side

$$f(t,x,\lambda) = \begin{pmatrix} \sigma_\lambda(t)(x_2 - x_1) \\ \rho_\lambda(t)x_1 - x_2 - x_1 x_3 \\ -\beta_\lambda(t)x_3 + x_1 x_2 \end{pmatrix}$$

in $\mathcal{D} = \mathbb{R} \times \mathbb{R}^3$. Here the coefficients $\sigma_\lambda, \rho_\lambda, \beta_\lambda : \mathbb{R} \to (0, \infty)$ are given by

$$\sigma_\lambda(t) = \sigma_0 + \lambda\sigma(t), \qquad \rho_\lambda(t) = 1 + \rho_0 + \lambda\rho(t), \qquad \beta_\lambda(t) = \beta_0 + \lambda\beta(t)$$

with real constants $\sigma_0, \beta_0 > 0$, $\rho_0 \in \mathbb{R}$, bounded C^3-functions σ, ρ, β and $\lambda \in \mathbb{R}$, which will serve as bifurcation parameter. It is our goal to study the stability of the equilibrium $x = 0$ of (4.4) for different values of λ. From the linearisation

$$D_2 f(t, 0, 0) = \begin{pmatrix} -\sigma_0 & \sigma_0 & 0 \\ 1 + \rho_0 & -1 & 0 \\ 0 & 0 & -\beta_0 \end{pmatrix},$$

we get that in case $\lambda = 0$, the trivial solution is uniformly asymptotically stable for $\rho_0 \in \left[-\left(\frac{\sigma_0 + 1}{2\sigma_0} \right)^2, 0 \right)$ and unstable for $\rho_0 > 0$. More interesting is the nonhyperbolic case $\rho_0 = 0$, where a pitchfork bifurcation occurs as ρ_0 passes through 0. To mimic this situation, we assume $\rho_0 = 0$ from now on. Before proceeding, we augment the original system (4.4) by λ as additional state space variable satisfying $\dot{\lambda} = 0$ and—to simplify our calculations—apply the linear transformation

$$\begin{pmatrix} y_1 \\ y_2 \\ y_3 \\ y_4 \end{pmatrix} := \begin{pmatrix} -\sigma_0 & 0 & 1 & 0 \\ 1 & 0 & 1 & 0 \\ 0 & 1 & 0 & 0 \\ 0 & 0 & 0 & 1 \end{pmatrix} \begin{pmatrix} x_1 \\ x_2 \\ x_3 \\ \lambda \end{pmatrix}$$

to the resulting equation $(\bar{D}_\lambda)$. This implies the system

$$\dot{y} = Ay + F(t, y) \tag{4.5}$$

in $\mathbb{R}^4$ with $A := \operatorname{diag}(-\sigma_0 - 1, -\beta_0, 0, 0)$ and the nonlinearity

$$F(t, y) := \begin{pmatrix} \frac{\sigma_0}{\sigma_0 + 1} y_1 y_2 - \frac{\sigma(t) + \sigma_0(\sigma(t) + \rho(t))}{\sigma_0 + 1} y_1 y_4 - \frac{1}{\sigma_0 + 1} y_2 y_3 + \frac{\rho(t)}{\sigma_0 + 1} y_3 y_4 \\ -\sigma_0 y_1^2 + (1 - \sigma_0) y_1 y_3 - \beta(t) y_2 y_4 + y_3^2 \\ \frac{\sigma_0^2}{\sigma_0 + 1} y_1 y_2 + \frac{\sigma(t) + \sigma_0(\sigma(t) - \sigma_0 \rho(t))}{\sigma_0 + 1} y_1 y_4 - \frac{\sigma_0}{\sigma_0 + 1} y_2 y_3 + \frac{\sigma_0 \rho(t)}{\sigma_0 + 1} y_3 y_4 \\ 0 \end{pmatrix}.$$

Thus, we can apply the centre manifold theorem [194, Theorem 3.2] yielding a centre manifold $\mathcal{C} \subseteq \mathbb{R} \times \mathbb{R}^3$ with two-dimensional fibres. The ansatz

$$c^-(t, y_3, y_4) = \sum_{i=0}^{2} y_3^{2-i} y_4^i \begin{pmatrix} c_{2-i,i}^1(t) \\ c_{2-i,i}^2(t) \end{pmatrix} + O\left(\sqrt{y_3^2 + y_4^2}^{\,3} \right)$$

shows that Eq. (4.5) reduced to the centre manifold $\mathcal{C}$ is given by

$$\dot{y}_3 = \frac{\sigma_0}{\sigma_0 + 1} \left(\lambda \rho(t) y_3 - c_{2,0}^2(t) y_3^3 \right) + O(\lambda y_3^2, \lambda^2 y_3, y_3^4).$$

Using (4.3), we obtain $c_{2,0}^2(t) \equiv \frac{1}{\beta_0}$, and consequently, the bifurcation equation is

$$\dot{y}_3 = \frac{\sigma_0}{\sigma_0 + 1} \left(\lambda \rho(t) y_3 - \frac{1}{\beta_0} y_3^3 \right) + r(t, y_3, \lambda), \tag{4.6}$$

where the remainder r satisfies the three limit relations

$$\lim_{y_3 \to 0} \sup_{\lambda \in (-|y_3|^3, |y_3|^3)} \sup_{t \in \mathbb{R}} \frac{|r(t, y_3, \lambda)|}{|y_3|^3} = 0,$$

$$\lim_{\lambda \to 0} \sup_{y_3 \in (-|\lambda|, |\lambda|)} \sup_{t \in \mathbb{R}} \frac{|r(t, y_3, \lambda)|}{|\lambda|^2} = 0,$$

$$\lim_{\lambda \to 0} \frac{1}{\lambda} \limsup_{y_3 \to 0} \sup_{t \in \mathbb{R}} \frac{|r(t, y_3, \lambda)|}{|y_3|} = 0.$$

We obtain a nonautonomous pitchfork bifurcation scenario from *[195, Proposition 5.2], but also [152, Theorem 8] applies. There exists a $\lambda_0 > 0$ such that:*

(i) *Supercritical case. If $\liminf_{t \to -\infty} \rho(t) > 0$, then for all $\lambda \in (0, \lambda_0)$ the trivial solution $\mathbb{R} \times \{0\}$ is a local repeller. Moreover, there exists a nontrivial local pullback attractor $\mathcal{A}_\lambda$ of (4.6) containing two bounded solutions ϕ_λ^- (pullback attracting in $(-\varepsilon, 0)$) and ϕ_λ^+ (pullback attracting in $(0, \varepsilon)$) for some $\varepsilon > 0$, and we have*

$$\lim_{\lambda \searrow 0} d(\mathcal{A}_\lambda(t), \{0\}) = 0 \quad \text{for all } t \in \mathbb{R}.$$

For all $\lambda \in (-\lambda_0, 0]$, the set $\mathbb{R} \times \{0\}$ is a local pullback attractor of (4.6).

(ii) *Subcritical case. If $\limsup_{t \to -\infty} \rho(t) < 0$, then for all $\lambda \in (-\lambda_0, 0)$ there exists a nontrivial local pullback attractor $\mathcal{A}_\lambda$ of (4.6), and we have*

$$\lim_{\lambda \nearrow 0} d(\mathcal{A}_\lambda(t), \{0\}) = 0 \quad \text{for all } t \in \mathbb{R}.$$

For $\lambda \in [0, \lambda_0)$, the set $\mathbb{R} \times \{0\}$ is a local pullback attractor of (4.6).

Not only the bifurcation equation (4.6) admits a bifurcation of this type but also the nonautonomous Lorenz equation (4.4) itself. This is due to an asymptotic phase property of the centre manifold (see [14, Theorem 4] for a global version). That is, every solution of (4.4) in a neighbourhood of the manifold approaches exponentially a solution on the centre manifold in forward time. Thus, for small $\lambda > 0$, there exists a local pullback attractor of (4.4) in $\mathbb{R} \times \mathbb{R}^d$ shrinking down to $\mathbb{R} \times \{0\}$ for $\lambda \searrow 0$.

We note that the criteria from [173, Section 5] impose global conditions on the bifurcation equation (4.6). Thus, in order to apply them, an ambient global extension of (4.6) is needed.

4.2 Normal Forms

A normal form theory for nonautonomous systems (D) near entire reference solutions $\phi^* : \mathbb{R} \to \mathbb{R}^d$ was developed in [216]. Our presentation requires continuous right hand sides $f : \mathcal{D} \to \mathbb{R}^d$ under the following assumptions:

(i) The variational equation (V) has bounded growth and a dichotomy spectrum $\Sigma(\phi^*)$ as in (4.1).

(ii) The derivatives $D_2^j f : \mathcal{D} \to L_j(\mathbb{R}^d)$ for $0 \leq j \leq m$ exist and are continuous.

(iii) $\sup_{t \in \mathbb{R}} \left\| D^j f(t, \phi^*(t)) \right\| < \infty$ for all $2 \leq j \leq m$.

This allows to establish a local C^m-equivalence between the entire solution ϕ^* of (D) and the trivial solution of nonautonomous differential equation

$$\dot{x} = B(t)x + G(t, x), \qquad (4.7)$$

being simplified in the following sense:

(a) The linear part $\dot{y} = B(t)y$ is in block diagonal form

$$B(t) := \begin{pmatrix} B_1(t) & & & \\ & B_2(t) & & \\ & & \ddots & \\ & & & B_n(t) \end{pmatrix} \in \mathbb{R}^{d \times d} \quad \text{for all } t \in \mathbb{R}$$

consisting of n continuous blocks $B_j : \mathbb{R} \to \mathbb{R}^{d_j \times d_j}$ with $\Sigma(B_j) = [a_j, b_j]$ for $1 \leq j \leq n$ and $d_1 + \ldots + d_n = d$. In particular, we have $\Sigma(\phi^*) = \Sigma(B)$.

(b) The right hand side of (4.7) is of class C^m in the second variable on a neighbourhood $B_{r_0}(0) \subset \mathbb{R}^d$ and thus allows the Taylor expansion

$$G(t, x) = \sum_{j=2}^{m} \sum_{|\alpha|=j} \frac{1}{\alpha!} D_2^\alpha G(t, 0) x^\alpha + R_m(t, x)$$

having a remainder satisfying $\lim_{x \to 0} \frac{1}{\|x\|^m} R_m(t, x) = 0$ uniformly in $t \in \mathbb{R}$. In this notation, $\alpha = (\alpha_1, \ldots, \alpha_n) \in \mathbb{N}_0^n$ denotes a *multi-index* of *length* $|\alpha| := \alpha_1 + \cdots + \alpha_n$ and $\alpha! := \alpha_1! \cdots \alpha_n!$.

The following 'interval arithmetic' is needed before stating the central result: for real numbers $a \leq b$, $c \leq d$, we define the sum $[a, b] + [c, d] := [a + c, b + d]$ of intervals and correspondingly the multiples $j[a, b] := [ja, jb]$ for $j \in \mathbb{N}$. The following can be found in [216].

Theorem 4.2.1 (Normal Form Theorem) *If* $1 \leq j \leq n$ *and a multi-index* $\alpha \in \mathbb{N}_0^n$ *with* $2 \leq |\alpha| \leq m$ *satisfies the* nonresonance condition

$$[a_j, b_j] \cap \sum_{i=1}^{n} \alpha_i [a_i, b_i] = \emptyset,$$

then $D_2^\alpha \Pi_j G(t, 0) \equiv 0$ *on* $\mathbb{R}$ *with the projector* $\Pi_j \in \mathbb{R}^{d \times d}$ *onto* $\{0\} \times \mathbb{R}^{d_j} \times \{0\}$.

Remarks

Integral Manifolds The extension of invariant manifolds to time-variant differential equations is called *integral manifolds*. A well-written and comprehensive introduction to integral manifolds is [20]; see [195] for their differentiability properties. The topic is also covered in [139, pp. 111ff, Chapter 6].

The reduction to centre manifolds in a nonautonomous setting dates back to [14]. By construction, the reduced equation (4.2) has a critical linear part meaning that its spectrum is an interval containing 0. Hence, unless (4.2) is scalar, corresponding stability investigations require subtle techniques.

Centre manifolds remain important in the more general nonhyperbolic situation shown in Fig. 4.2b: a general spectral interval $[a_{i^*}, b_{i^*}]$ and not necessarily the dominant one contains 0. Our above approach yields a *centre-unstable manifold*, now denoted as $\mathcal{C}^-$, and the subsequent spectral gap to the right between b_{i^*} and a_{i^*+1}, $1 \leq i^* < n$, allows to construct a *centre-stable manifold* $\mathcal{C}^+$ on the interval $\mathbb{I} = \mathbb{R}$. Then the intersection $\mathcal{C} := \mathcal{C}^+ \cap \mathcal{C}^-$ forms a *centre manifold* $\mathcal{C}$; see [20]. This nonautonomous set still contains all solutions staying near the reference solution ϕ^*. It is rather a saddle and not necessarily attractive anymore [21]. Whence, $\mathcal{C}$ still contains bifurcating solutions, but one can speak of bifurcating attractors (repellers) only when the differential equation is restricted to $\mathcal{C}^-$ (respectively, $\mathcal{C}^+$).

Finally, to approximate integral manifolds analytically, we refer to [195]. An effective algorithm to obtain graphical illustration is given in [109].

Normal Forms An algebraic approach to simplify differential equations is normal forms. The corresponding nonautonomous theory has been established in [216] in the general set-up of Carathéodory equations. It provides a nice and natural formulation of the nonresonance conditions in terms of the dichotomy spectrum containing the classical autonomous situation as a special case.

Concerning a normal form theory for quasi-periodic equations, we refer to [49].

Part II
Nonautonomous Difference Equations

The second part is devoted to discrete time $\mathbb{T} = \mathbb{Z}$ and difference equations. Let $\mathbb{I}$ denote the intersection of a real interval with the integers $\mathbb{Z}$, a so-called *discrete interval*, and define the *lapped interval* $\mathbb{I}' := \{t \in \mathbb{I} : t + 1 \in \mathbb{I}\}$. Typically, $\mathbb{I}$ will be one of the axes $\mathbb{Z}_0^-$, $\mathbb{Z}_0^+$ or $\mathbb{Z}$. Certain concretisations in the present situation motivate us to consider processes and skew products separately:

Discrete Processes We write nonautonomous difference equations [72, 186] as

$$x_{t+1} = f(t, x_t) = f_t(x_t) \tag{Δ}$$

with a continuous right hand side $f : \mathcal{D} \to \mathbb{R}^d$ on an open nonautonomous set $\mathcal{D} \subseteq \mathbb{I} \times \mathbb{R}^d$ and $f_t := f(t, \cdot)$. The forward solution to (Δ) satisfying the initial condition $x_\tau = \xi$ for given initial pairs $(\tau, \xi) \in \mathcal{D}$ is called *general solution*; it is denoted by $\varphi(\cdot, \tau, \xi)$. In contrast to differential equations, forward existence is not a problem and one has

$$\varphi(t, \tau, \xi) = \begin{cases} f_{t-1} \circ \ldots \circ f_\tau(\xi), & \tau < t, \\ \xi, & t = \tau, \end{cases}$$

as long as the iterates stay in $\mathcal{D}(t)$. If $f_t(\mathcal{D}(t)) \subseteq \mathcal{D}(t+1)$ holds for all $t \in \mathbb{I}'$, then φ is a process.

Certain notions simplify for difference equations. For instance, a nonautonomous set $\mathcal{A} \subseteq \mathcal{D}$ is *invariant*, if and only $\mathcal{A}(t + 1) = f_t(\mathcal{A}(t))$ for all $t \in \mathbb{I}'$ and *forward invariant*, if the inclusion $\mathcal{A}(t + 1) \subseteq f_t(\mathcal{A}(t))$ holds for every $t \in \mathbb{I}'$, which in turn is equivalent to $\mathcal{A}(t) = \varphi(t, \tau, \mathcal{A}(\tau))$ (see Fig. 1.1), respectively, the inclusion $\mathcal{A}(t) \subseteq \varphi(t, \tau, \mathcal{A}(\tau))$ for all $\tau \leq t$.

In general, difference equations are easier to handle concerning the existence of forward solutions, but results on the backward behaviour often require additional assumptions. Without invertibility of f_t, backward solutions to (Δ) might not exist or might not be unique. For bijective $f_t : \mathcal{D}(t) \to \mathcal{D}(t+1)$ with continuous inverses f_t^{-1}, in order to obtain an invertible process, one additionally sets

$$\varphi(t, \tau, \cdot) := f_t^{-1} \circ \ldots \circ f_{\tau-1}^{-1} \quad \text{for all } t < \tau.$$

In order to tackle linear equations, assume that $A_t \in \mathbb{R}^{d \times d}$, $t \in \mathbb{I}'$, is a bounded sequence of *coefficient matrices* and consider a linear difference equation

$$x_{t+1} = A_t x_t \tag{Λ}$$

on the extended state space $\mathcal{D} = \mathbb{I} \times \mathbb{R}^d$. The *transition matrix* is explicitly given as

$$\Phi(t, s) := \begin{cases} A_{t-1} \cdots A_s, & s < t, \\ \mathrm{id}, & t = s, \end{cases}$$

and, if the matrices A_t are invertible, we set $\Phi(t, s) := A_t^{-1} \cdots A_{s-1}^{-1}$ for $t < s$.

Bifurcation theory deals with parametrised equations

$$x_{t+1} = f_t(x_t, \lambda) \qquad\qquad (\Delta_\lambda)$$

with λ from some parameter space Λ. The associated general solution (process) will be denoted as φ_λ. It is as smooth as the right hand side of (Δ_λ) in the variables ξ, λ.

Discrete Skew Product Flows Also the skew product formulation of driven difference equations

$$x_{t+1} = f(\theta(t, \omega), x_t) \qquad\qquad (\Delta^\omega)$$

with a continuous right hand side $f : \Omega \times \mathbb{R}^d \to \mathbb{R}^d$ becomes more concrete in discrete time. First of all, the base flow $\theta : \mathbb{Z} \times \Omega \to \Omega$ on a metric space Ω is always of the form $\theta(t, \omega) := \theta_0^t(\omega)$, where θ_0^t denote the iterates of a homeomorphism $\theta_0 := \theta(1, \cdot)$ on Ω. Second, the *cocycle*, i.e. the solution starting at time 0 in $\xi \in \mathbb{R}^d$, has the explicit representation

$$\varphi(t, \omega, \xi) = f(\theta(t-1, \omega), \cdot) \circ \ldots \circ f(\theta(1, \omega), \cdot) \circ f(\omega, \xi) \quad \text{for all } t \in \mathbb{Z}_0^+, \omega \in \Omega.$$

Motivated by Fig. 1.2, φ is also called *fibre map*.

Given this, the invariant subsets $\mathcal{A} \subseteq \Omega \times \mathbb{R}^d$ can be characterised by means of the property $\mathcal{A}(\theta_0(\omega)) = f(\omega, \mathcal{A}(\omega))$ for all $\omega \in \Omega$.

In case of driven linear equations

$$x_{t+1} = A(\theta(t, \omega))x_t \qquad\qquad (\Lambda^\omega)$$

with coefficients $A : \Omega \to \mathbb{R}^{d \times d}$, their *linear cocycle* reads explicitly as

$$\Phi(t, \omega) := A(\theta(t-1, \omega)) \cdots A(\theta(1, \omega))A(\omega) \quad \text{for all } t \in \mathbb{Z}_0^+, \omega \in \Omega.$$

Chapter 5
Spectral Theory, Stability and Continuation

Problems parallel to those of Chap. 2 also arise for linear equations in discrete time. Among them, eigenvalues and other notions from classical linear algebra [53, 102] lose their relevance for time-dependent problems. More suitable approaches to a "nonautonomous linear algebra" provide the Spectral Theorem (based on exponential dichotomies) when dealing with linear processes and the Multiplicative Ergodic Theorem (based on Lyapunov exponents) suitable for linear cocycles. Variants of both results are formulated in this chapter.

Note that due to non-invertibility reasons, the difference equations case is not completely analogous to the theory presented in Sect. 2.1—besides the fact that the stability boundary is now shifted from 0 to 1. We, moreover, provide a more comprehensive discussion. It includes a rather detailed insight into different forms of nonhyperbolicity. For instance, the fine structure of the dichotomy spectrum allows to classify nonautonomous bifurcations on a linear basis already. In addition, further information on Lyapunov exponents when dealing with discrete linear skew product flows is given. We formulate the Multiplicative Ergodic Theorem for one-sided, noninvertible systems.

The section addressing stability theory is largely analogous to the differential equations case, but this time a sufficient condition for exponential stability in the framework of the Multiplicative Ergodic Theorem is given.

The section on continuation again emphasises the importance of exponential dichotomies. On $\mathbb{Z}$ they yield continuation results. On half lines they form the basis of the functional analytical approach to bifurcations given in Sect. 6.2. We thus elaborate on the non-uniqueness problems related to the continuation of solutions on half lines.

5.1 Spectral Theory

The aim of this section is to provide appropriate dynamical spectra for nonautonomous linear difference equations of the form (Λ).

© The Author(s), under exclusive license to Springer Nature Switzerland AG 2023
V. Anagnostopoulou et al., *Nonautonomous Bifurcation Theory*, Frontiers
in Applied Dynamical Systems: Reviews and Tutorials 10,
https://doi.org/10.1007/978-3-031-29842-4_5

We have already demonstrated in continuous time that the eigenvalues of the coefficient matrices for linear nonautonomous differential equations do not contain information on stability, unless there is a particularly slow time-dependence. The same also holds for nonautonomous difference equations (Λ), and a corresponding example in form of a periodic difference equation can be found in [72, Example 4.17]. Although this example is periodic, the spectrum of a periodic difference equation (Λ) is well understood using Floquet theory. If the difference equation (Λ) is T-periodic (i.e. $A_{t+T} = A_t$ for all $t \in \mathbb{Z}$), then the so-called *monodromy matrix* $\Pi := \Phi(T, 0) = A_{T-1} \cdots A_0$ can be computed explicitly (note that this is normally not possible in the continuous-time case). The eigenvalues of the monodromy matrix are called *Floquet multipliers* of (Λ), and they allow application of the classical stability and perturbation theory.

The situation in the non-periodic case is more subtle, as in the differential equations case, and in this section, we discuss two types of spectral notions: discrete time extensions of the dichotomy spectrum (first studied by R.J. Sacker and G.R. Sell) and the Lyapunov spectrum (due to A.M. Lyapunov), see also [17, 207] and [24].

5.1.1 Dichotomy Spectrum

Given the fact that asymptotic stability is not a robust stability notion (see [158, §2] and [186, Example 3.4.1]), one could argue that a more feasible concept is uniform exponential stability. Its natural generalisation is an exponential dichotomy [99, p. 229, Definition 7.6.4], which we discuss next.

Consider the linear difference equation (Λ) on an unbounded discrete interval $\mathbb{I}$. An *invariant projector* for (Λ) is a sequence $(P_t)_{t \in \mathbb{I}}$ in $\mathbb{R}^{d \times d}$ of projections $P_t = P_t^2$ such that $A_{t+1}P_t = P_t A_t$ and

$$A_t|_{N(P_t)} : N(P_t) \to N(P_{t+1}) \quad \text{is invertible for all } t \in \mathbb{I}'. \tag{5.1}$$

This guarantees that for all $t \geq s$, the restriction $\Phi(t, s)|_{N(P_s)} : N(P_s) \to N(P_t)$ is well defined and invertible with inverse denoted by $\Phi(s, t)$. Moreover, it ensures that all the kernels $N(P_t)$ for $t \in \mathbb{I}$ have the same dimension.

Definition 5.1.1 (Exponential Dichotomy) *The linear difference equation (Λ) is said to admit an* exponential dichotomy *on $\mathbb{I}$ if there exist an invariant projector $(P_t)_{t \in \mathbb{I}}$ and real numbers $K \geq 1$ and $\beta \in (0, 1)$ such that*

$$\|\Phi(t, s)P_s\| \leq K\beta^{t-s}, \qquad \|\Phi(s, t)(\mathrm{id} - P_t)\| \leq K\beta^{t-s} \quad \text{for all } s \leq t.$$

The *stable nonautonomous set* $\mathcal{V}^+$ of (Λ) is given as in Sect. 2.1.1, while noninvertibility requires to define the *unstable nonautonomous set* differently as

$$\mathcal{V}^- := \left\{ (\tau, \xi) \in \mathbb{I} \times \mathbb{R}^d : \begin{array}{l} \text{there exists a solution } \phi = (\phi_t)_{t \in \mathbb{I}} \\ \text{with } \phi_\tau = \xi \text{ and } \lim_{t \to -\infty} \phi_t = 0 \end{array} \right\},$$

where we assume that in this case, $\mathbb{I}$ is unbounded below (see also Fig. 5.1).

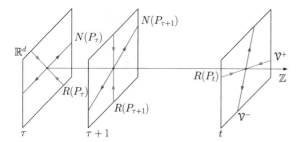

Figure 5.1 Fibres for the stable set $\mathcal{V}^+(t) = R(P_t)$ (green) and the unstable set $\mathcal{V}^-(t) = N(P_t)$, $t \in \mathbb{I}$ (red), to a linear difference equation (Λ) with an exponential dichotomy. They intersect along the trivial solution

The following proposition taken from [15, Theorem 2.5] explains how the stable and unstable nonautonomous sets relate to an invariant projector.

Proposition 5.1.2 *Assume that the linear system (Λ) has an exponential dichotomy on $\mathbb{I}$ with an invariant projector $(P_t)_{t \in \mathbb{I}}$. If $\mathbb{I}$ is unbounded above, then $R(P_t) = \mathcal{V}^+(t)$ for all $t \in \mathbb{I}$, and the range of P_t is thus uniquely determined. For $\mathbb{I}$ unbounded below, the unstable nonautonomous set $\mathcal{V}^-$ allows the dynamical characterisation $N(P_t) = \mathcal{V}^-(t)$ for all $t \in \mathbb{I}$.*

Note that similar to the continuous-time case, a Roughness Theorem holds for exponential dichotomies also in discrete time (see [99, p. 232, Theorem 7.6.7] or [191, p. 165, Theorem 3.6.5]). This means that an exponential dichotomy is an open property in the class of linear difference equations (Λ) with bounded coefficient sequences $(A_t)_{t \in \mathbb{I}'}$. Nonetheless, [191, p. 149, Example 3.4.34] shows that an exponential dichotomy is not generic for general time-dependencies, as opposed to the autonomous and periodic situations.

The following counterpart to Theorem 2.1.4 links an exponential dichotomy on $\mathbb{Z}$ to the exponential dichotomies on both half lines (see [31, Lemma 2.4] or [25, Corollary 2] for the noninvertible case).

Theorem 5.1.3 (Characterisation of Exponential Dichotomies) *The linear system (Λ) admits an exponential dichotomy on $\mathbb{Z}$ if and only if the following two conditions are fulfilled:*

(a) *(Λ) has an exponential dichotomy on $\mathbb{Z}_0^+$ with projector P_t^+ and an exponential dichotomy on $\mathbb{Z}_0^-$ with projector P_t^-.*

(b) *$R(P_0^+) \oplus N(P_0^-) = \mathbb{R}^d$.*

We introduce the corresponding discrete-time version of the dichotomy spectrum from Definition 2.1.5. Consider the scaled difference equation

$$x_{t+1} = \gamma^{-1} A_t x_t , \qquad\qquad (\Lambda_\gamma)$$

and we will be interested for which real $\gamma > 0$, the scaled system (Λ_γ) admits an exponential dichotomy.

Definition 5.1.4 (Dichotomy Spectrum) *The* dichotomy spectrum *of* (Λ) *on the unbounded discrete interval* $\mathbb{I}$ *is defined by*

$$\Sigma_{\mathbb{I}}(A) := \{\gamma > 0 : (\Lambda_\gamma) \text{ does not have an exponential dichotomy on } \mathbb{I}\}.$$

Among the special cases

$$\Sigma(A) := \Sigma_{\mathbb{Z}}(A), \qquad \Sigma^+(A) := \Sigma_{\mathbb{Z}_0^+}(A), \qquad \Sigma^-(A) := \Sigma_{\mathbb{Z}_0^-}(A),$$

the latter two are called forward *and* backward dichotomy spectrum *of* (Λ), *respectively, and for simplicity we often refer to them as* dichotomy spectra.

We point out an elegant connection to operator theory. The linear system (Λ_γ) admits an exponential dichotomy on $\mathbb{Z}$ if and only if the linear operator

$$S_\gamma : \ell^\infty(\mathbb{R}^d) \to \ell^\infty(\mathbb{R}^d), \qquad (S_\gamma \phi)_t := \phi_{t+1} - \gamma^{-1} A_t \phi_t \quad \text{for all } t \in \mathbb{Z}$$

has a bounded inverse, i.e. $S_\gamma \in GL(\ell^\infty(\mathbb{R}^d))$, see [99, p. 230, Theorem 7.6.5]). On this basis, the characterisation $\Sigma(A) = \{\gamma > 0 : S_\gamma \notin GL(\ell^\infty(\mathbb{R}^d))\}$ holds.

We note that, contrary to the continuous-time case, super- and sub-exponential growth is not measured by the dichotomy spectrum from Definition 5.1.4. Hence, the extremal growth rates 0 and ∞ cannot be part of the spectrum, and this has consequences for the formulation of the spectral theorem, see [17, 27, 28, 207].

Theorem 5.1.5 (Spectral Theorem) *Consider the linear difference equation* (Λ) *on the unbounded discrete interval* $\mathbb{I}$ *with a bounded coefficient sequence* $(A_t)_{t \in \mathbb{I}'}$. *Then the dichotomy spectrum* $\Sigma_{\mathbb{I}}(A)$ *of* (Λ) *is the disjoint union of* $n \leq d$ *nonempty spectral intervals* $\sigma_1, \ldots, \sigma_n \subseteq (0, \infty)$, *where*

$$\sigma_1 = \begin{cases} [a_1, b_1] \\ or \\ (0, b_1], \end{cases} \qquad \sigma_i = [a_i, b_i] \quad \text{for all } 2 \leq i \leq n,$$

with reals $0 < a_1 \leq b_1 < a_2 \leq \ldots < b_n$. *We call* $\sigma_n = [a_n, b_n]$ *the* dominant spectral interval, *and the additional assumption* $A_t \in GL(\mathbb{R}^d)$ *for all* $t \in \mathbb{I}'$ *with* $\sup_{t \in \mathbb{I}'} \|A_t^{-1}\| < \infty$ *ensures that* $\sigma_1 = [a_1, b_1]$.

A necessary and sufficient condition guaranteeing that all solutions of (Λ) grow and decay at most exponentially, that is there exist $K \geq 1$ and $\alpha > 1$ such that

$$\|\Phi(t, s)\| \leq K \alpha^{|t-s|} \quad \text{for all } t, s \in \mathbb{I},$$

is invertibility $A_t \in GL(\mathbb{R}^d)$ combined with $\sup_{t \in \mathbb{I}'} \max\{\|A_t\|, \|A_t^{-1}\|\} < \infty$. In analogy to the continuous-time situation from Sect. 2.1, such difference equations (Λ) are said to have *bounded growth*. They satisfy $\Sigma(A) \subseteq [\frac{1}{\alpha}, \alpha]$, see [16].

We now aim at constructing invariant nonautonomous sets corresponding to the spectral intervals under the assumption that $\mathbb{I} = \mathbb{Z}$. First, given a growth rate $\gamma > 0$, we define the

- γ-*stable nonautonomous set*

$$\mathcal{V}_\gamma^+ := \left\{ (\tau, \xi) \in \mathbb{Z} \times \mathbb{R}^d : \sup_{\tau \leq t} |\Phi(t, \tau)\xi| \, \gamma^{\tau-t} < \infty \right\}$$

- The γ-*unstable nonautonomous set*

$$\mathcal{V}_\gamma^- := \left\{ (\tau, \xi) \in \mathbb{Z} \times \mathbb{R}^d : \begin{array}{l} \text{there exists a solution } \phi = (\phi_t)_{t \in \mathbb{Z}} \text{ with} \\ \phi_\tau = \xi \text{ and } \sup_{t \leq \tau} \|\phi_t\| \, \gamma^{\tau-t} < \infty \end{array} \right\},$$

whose fibres are linear subspaces of $\mathbb{R}^d$. In particular, $\mathcal{V}^\pm = \mathcal{V}_1^\pm$. Choosing the rates $\gamma_i \in (b_i, a_{i+1})$ for $1 \leq i < n$ and $\gamma_0 \in (0, a_1)$ whenever $\sigma_1 = [a_1, b_1]$, we define

$$\mathcal{V}_0 := \begin{cases} \mathcal{V}_{\gamma_0}^+ & : \quad \sigma_1 = [a_1, b_1], \\ \mathbb{Z} \times \{0\} & : \quad \sigma_1 = (0, b_1], \end{cases}$$

$$\mathcal{V}_1 := \mathcal{V}_{\gamma_1}^+ \cap \begin{cases} \mathcal{V}_{\gamma_0}^- & : \quad \sigma_1 = [a_1, b_1], \\ \mathbb{Z} \times \mathbb{R}^d & : \quad \sigma_1 = (0, b_1], \end{cases}$$

$$\mathcal{V}_i := \mathcal{V}_{\gamma_i}^+ \cap \mathcal{V}_{\gamma_{i-1}}^- \quad \text{for all } i \in \{2, \ldots, n\} .$$

The vector bundles $\mathcal{V}_0, \ldots, \mathcal{V}_n$ are forward invariant with fibres of constant dimension $\dim \mathcal{V}_i$, called *multiplicity* of the corresponding spectral interval σ_i for indices $1 \leq i \leq n$ and the *Whitney sum* $\mathbb{Z} \times \mathbb{R}^d = \mathcal{V}_0 \oplus \ldots \oplus \mathcal{V}_n$. Note that in case $\mathcal{V}_0$ is nontrivial, then we have sub-exponential growth, and solutions with such growth behaviour lie in this nonautonomous set.

We now provide dichotomy spectra of the linear system (Λ) in special cases:

- *One-dimensional case.* For invertible scalar difference equations $x_{t+1} = a_t x_t$ satisfying $\sup_{t \in \mathbb{I}'} \left\{ |a_t|, |a_t^{-1}| \right\} < \infty$, the spectrum is related to Bohl exponents in terms of $\Sigma_{\mathbb{I}}(a) = [\underline{\beta}_{\mathbb{I}}(a), \overline{\beta}_{\mathbb{I}}(a)]$ [27, 185, Theorem 4.6], see Appendix A.2 for the definition of the *Bohl exponents*. In the asymptotically constant case $a_t = a^+$ for $t \geq \tau^+$ and $a_t = a^-$ for $t \leq \tau^-$, where $a^+, a^- \neq 0$, one obtains $\underline{\beta}_{\mathbb{I}}(a) = \min \left\{ |a^+|, |a^-| \right\}$ and $\overline{\beta}_{\mathbb{I}}(a) = \max \left\{ |a^+|, |a^-| \right\}$.

- *Autonomous case.* Similarly to continuous time, the dichotomy spectra for an autonomous linear difference equation $x_{t+1} = A x_t$ relate to the eigenvalues of A, here to the absolute values of the eigenvalues. The dichotomy spectra are given by $\Sigma(A) = \Sigma^+(A) = \Sigma^-(A) = |\sigma(A)| \setminus \{0\}$.

- *T-periodic case.* For a T-periodic equation (Λ) with the monodromy matrix $\Pi = \Phi(T, 0)$, one obtains that $\Sigma(A) = \Sigma^+(A) = \Sigma^-(A) = \sqrt[T]{|\sigma(\Pi)|} \setminus \{0\}$ [27, Theorem 4.1].

- *Asymptotically autonomous case.* If the coefficient sequence $A_t \in GL(\mathbb{R}^d)$ in Eq. (Λ) is *asymptotically autonomous*, that is

$$A^+ := \lim_{t \to \infty} A_t, \qquad\qquad A^- := \lim_{t \to -\infty} A_t$$

with invertible limits $A^+, A^- \in \mathbb{R}^{d \times d}$, then one obtains the dichotomy spectra

$$\Sigma^+(A) = \left|\sigma(A^+)\right| \setminus \{0\}, \quad \Sigma^-(A) = \left|\sigma(A^-)\right| \setminus \{0\} \quad \text{for all } \tau \in \mathbb{Z}.$$

To determine the spectrum, $\Sigma(A)$ is more involved [27, Theorem 4.8].

Remark 5.1.6 (Dependence of the Dichotomy Spectra on Perturbations) *The dichotomy spectra $\Sigma^\pm(A)$ and $\Sigma(A)$ depend upper semi-continuously on perturbations of $(A_t)_{t \in \mathbb{I}'}$ in the ℓ^∞-topology, see [185, Corollary 4] and [192, Corollary 3.24]). Hence, it is difficult to establish a smooth perturbation theory for spectral intervals as it is possible for eigenvalues in the autonomous or periodic case. Nevertheless, the set of discontinuity points for the set-valued functions $\Sigma^\pm$, Σ is meagre [192, Remark 4.26(1)]. On the other hand, as long as one deals with invertible matrices, the dichotomy spectra $\Sigma^\pm(A)$ are invariant under ℓ_0-perturbations, i.e., one has the relation $\Sigma^\pm(A) = \Sigma^\pm(A + B)$ for matrix sequences $B_t \in \mathbb{R}^{d \times d}$ with $\lim_{t \to \pm\infty} B_t = 0$ [28, Theorem 2.3].*

The next example illustrates that the dichotomy spectra are not continuous for perturbations in the ℓ^∞-topology, and it shows that the above statement with regard to ℓ_0-perturbations fails for the dichotomy spectrum $\Sigma(A)$.

Example 5.1.7 (Dichotomy Spectrum Under Perturbations) *Suppose that δ and ε are reals with $0 < |\varepsilon| < 1 < |\delta|$. We consider a planar linear difference equation $x_{t+1} = A_t^\lambda x_t$ depending on a real parameter λ with*

$$A_t^\lambda := \begin{pmatrix} a_t & \lambda \varepsilon_t \\ 0 & a_t^{-1} \end{pmatrix}, \quad a_t := \begin{cases} \delta &: t \geq 0, \\ \frac{1}{\delta} &: t < 0, \end{cases} \quad \varepsilon_t := \begin{cases} \varepsilon^t &: t \geq 0, \\ 0 &: t < 0. \end{cases}$$

Using the above result on spectra for scalar linear equations, we obtain

$$\Sigma(A^0) = \left[\tfrac{1}{|\delta|}, |\delta| \right]$$

for the system where $\lambda = 0$, and we consider the matrix sequence $\lambda \begin{pmatrix} 0 & \varepsilon_t \\ 0 & 0 \end{pmatrix}$, $t \in \mathbb{Z}$, as perturbation of this system. Due to $\sup_{t \in \mathbb{Z}} |\lambda \varepsilon_t| = |\lambda|$, this perturbation can be made arbitrarily small; it decays to 0 exponentially, and we show now that it does affect the spectrum $\Sigma(A^\lambda)$. For $\gamma > 0$, the transition matrix Φ_γ of the scaled perturbed system (Λ_γ) satisfies

$$\Phi_\gamma(t, 0) = \gamma^{-t} \begin{pmatrix} \delta^t & \frac{\lambda\delta}{\delta^2 - \varepsilon}\left(\delta^t - \left(\frac{\varepsilon}{\delta}\right)^t\right) \\ 0 & \delta^{-t} \end{pmatrix} \quad \text{for all } t \in \mathbb{Z}_0^+,$$

whenever $\lambda \neq 0$. The γ-stable 0-fibre is given by

$$V_\gamma^+(0) = \begin{cases} \mathbb{R}^2 & : \quad |\delta| \leq \gamma, \\ \mathbb{R}\begin{pmatrix} \delta\lambda \\ \varepsilon-\delta^2 \end{pmatrix} & : \quad \frac{1}{|\delta|} \leq \gamma < |\delta|, \\ \{(0,0)\} & : \quad \gamma < \frac{1}{|\delta|} \end{cases}$$

and the γ-unstable 0-fibre reads as

$$V_\gamma^-(0) = \begin{cases} \{(0,0)\} & : \quad |\delta| < \gamma, \\ \{0\} \times \mathbb{R} & : \quad \frac{1}{|\delta|} < \gamma \leq |\delta|, \\ \mathbb{R}^2 & : \quad \gamma \leq \frac{1}{|\delta|}. \end{cases}$$

Hence, for values $\gamma \notin \{|\delta|, \frac{1}{|\delta|}\}$, we obtain the direct sum $V_\gamma^+(0) \oplus V_\gamma^-(0) = \mathbb{R}^2$ and Proposition 5.1.2 combined with Theorem 5.1.3 shows that (Λ_γ) admits an exponential dichotomy on the whole line $\mathbb{Z}$. This manifests a change in the dichotomy spectrum under the above perturbations, since we can conclude (Fig. 5.2)

$$\Sigma(A^\lambda) = \begin{cases} \{\frac{1}{|\delta|}, |\delta|\} & : \quad \lambda \neq 0, \\ [\frac{1}{|\delta|}, |\delta|] & : \quad \lambda = 0. \end{cases}$$

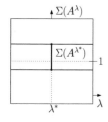

Figure 5.2 Dichotomy spectrum $\Sigma(A^\lambda)$ in Example 5.1.7 being discrete for parameters $\lambda \neq \lambda^* = 0$ and an interval containing 1 in the critical case $\lambda = \lambda^*$

We already pointed out that an exponential dichotomy of a linear difference equation (Λ_γ) on $\mathbb{Z}$ can be characterised in terms of invertibility of the shift operator with

$$S_\gamma : \ell^\infty(\mathbb{R}^d) \to \ell^\infty(\mathbb{R}^d), \qquad (S_\gamma\phi)_t := \phi_{t+1} - \gamma^{-1}A_t\phi_t.$$

In Sects. 6.1 and 6.2, it will be apparent that bifurcations in nonlinear problems (Δ_λ) can only occur in the absence of an exponential dichotomy. Therefore, it is crucial to investigate different forms of non-invertibility for S_γ. This gives rise to the following subsets of the dichotomy spectrum $\Sigma(A)$:

- The *point spectrum* $\Sigma_p(A) := \{\gamma > 0 : \dim S_\gamma^{-1}(\{0\}) > 0\}$
- The *surjectivity spectrum* $\Sigma_s(A) := \{\gamma > 0 : S_\gamma \text{ is not onto}\}$
- The *Fredholm spectra* $\Sigma_F(A) := \{\gamma > 0 : S_\gamma \text{ is not Fredholm}\}$, as well as $\Sigma_{F_0}(A) := \{\gamma > 0 : S_\gamma \text{ is not Fredholm or of nonzero index}\}$

Note that also the set-valued mappings $\Sigma_s, \Sigma_F, \Sigma_{F_0}$ are upper semi-continuous on the set of linear equations (Λ) with bounded coefficient sequences [192, Corollarys 4.21(c) and 4.26(c)]).

We will illustrate below these different dichotomy spectra allow a classification of nonautonomous bifurcations already on a linear level. Moreover, one can deduce the following relations between them [192, Corollary 4.31].

Theorem 5.1.8 (Fine Structure of the Dichotomy Spectrum) *If* $A_t \in GL(\mathbb{R}^d)$ *for all* $t \in \mathbb{Z}$, *then*

$$\Sigma_p(A) \subseteq \Sigma_p(A) \cup \Sigma_s(A)$$
$$\|$$
$$\Sigma^+(A) \cup \Sigma^-(A) = \Sigma_F(A) \subseteq \Sigma_{F_0}(A) \subseteq \quad \Sigma(A) \ = \Sigma_s(A) \cup \Sigma_{F_0}(A)$$
$$\cap\| \qquad\qquad\qquad\qquad\qquad \|$$
$$\partial\Sigma(A) \subseteq \Sigma_s(A) \qquad\qquad \subseteq \quad \Sigma_p(A) \cup \Sigma_{F_0}(A).$$

5.1.2 Lyapunov Spectrum

In this subsection, we discuss an alternative approach to obtain a 'nonautonomous linear algebra' suitable for linear difference equations, which is based on Lyapunov exponents and the related spectrum. While our presentation in the continuous time case in Sect. 2.1.2 was focussed on processes (see [24, pp. 56ff, § 2.5] for related discrete time results), we now address skew products.

We begin with a driving system $\theta : \mathbb{Z} \times \Omega \to \Omega$ on a metric space Ω, which is measure-preserving with respect to a probability space $(\Omega, \mathfrak{B}, \mu)$.

Suppose that $A : \Omega \to \mathbb{R}^{d \times d}$ is measurable and denote the linear cocycle generated by the difference equation

$$x_{t+1} = A(\theta(t, \omega))x_t \qquad\qquad (\Lambda^\omega)$$

as $\Phi(t, \omega) \in \mathbb{R}^{d \times d}$. The *Lyapunov exponent* of (Λ^ω) is given by

$$\chi(\omega, A) := \limsup_{t \to \infty} \sqrt[t]{\|\Phi(t, \omega)\|} \quad \text{for all } \omega \in \Omega. \qquad (5.2)$$

Under the assumptions of the subsequent result [84, Theorem 4.1], the above limit superior is actually a limit and constant for almost all $\omega \in \Omega$.

Theorem 5.1.9 (Multiplicative Ergodic Theorem) *Suppose that* $\theta : \mathbb{Z} \times \Omega \to \Omega$ *is ergodic with respect to the invariant probability measure* $\mu : \mathfrak{B} \to [0, 1]$. *If* $\ln^+ \|A(\cdot)\|$ *is integrable, then there exist reals* $0 \leq \chi_1 < \chi_2 < \ldots < \chi_k$ *and* $d_1, \ldots, d_k \in \mathbb{N}$ *satisfying* $d_1 + \ldots + d_k = d$, *as well as a measurable family of so-called* Oseledets *spaces* $V_1(\omega), \ldots, V_k(\omega) \subseteq \mathbb{R}^d$ *such that for almost every* $\omega \in \Omega$ *the following hold. The numbers* $\chi_1, \ldots, \chi_k$ *are Lyapunov exponents of* (Λ^ω), *i.e.*

$$\chi_i = \lim_{t \to \infty} \sqrt[t]{\|\Phi(t, \omega)\xi\|} \quad \text{for all } \xi \in V_i(\omega) \setminus \{0\},$$

the Oseledets spaces satisfy $A(\omega)V_i(\omega) \subseteq V_i(\theta(1,\omega))$ *with equality if* $\chi_i > 0$, $\dim V_i(\omega) = d_i$ *for all* $i \in \{1, \ldots, k\}$ *and one has the* Oseledets sum

$$V_1(\omega) \oplus \ldots \oplus V_k(\omega) = \mathbb{R}^d.$$

The finite set $\Sigma_{\mathrm{lyap}}(A) := \{\chi_i, \ldots, \chi_k\}$ is denoted as *Lyapunov spectrum* of (Λ^ω).

By means of arguments from the authors in [53, pp. 18ff, §1.5], who consider the case of invertible $A(\omega)$, the Multiplicative Ergodic Theorem 5.1.9 allows the following interpretation. The Lyapunov exponents χ_i are a nonautonomous counterpart to the eigenvalue moduli for autonomous problems $x_{t+1} = Ax_t$. Furthermore, the Oseledets spaces correspond to the sum of generalised eigenspaces for eigenvalues having the same moduli. Since the $V_i(\omega)$ have constant dimension, we denote $\dim V_i$ as the *multiplicity* of the corresponding Lyapunov exponent χ_i for $i \in \{1, \ldots, k\}$.

5.2 Stability

Let us first record the role of the dichotomy spectrum as hyperbolicity notion for nonautonomous difference equations (Δ) of class C^1. A solution $\phi^* = (\phi_t^*)_{t \in \mathbb{I}}$ of (Δ) is called *hyperbolic* (on $\mathbb{I}$), provided the variational equation

$$x_{t+1} = Df_t(\phi_t^*)x_t \qquad\qquad (Y)$$

has an exponential dichotomy on an unbounded discrete interval $\mathbb{I}$. If P_t denotes the corresponding invariant projector, then the constant dimension of $N(P_t)$, $t \in \mathbb{I}$, is called the *Morse index* of ϕ^* and indicates the number of unstable directions.

The relevance of the dichotomy spectrum $\Sigma_{\mathbb{I}}(\phi^*)$ of (Y) in stability theory is justified by the following result proven as Theorem A.1.1.

Proposition 5.2.1 (The Dichotomy Spectrum and Linearised Stability) *Consider the difference equation* (Δ) *with solution* ϕ^* *and corresponding variational equation* (Y). *If we assume that* $\mathbb{I}$ *is bounded below and*

$$\sup_{t \in \mathbb{I}} \|Df_t(\phi_t^*)\| < \infty, \qquad \lim_{x \to 0} \sup_{t \in \mathbb{I}} \|Df_t(\phi_t^* + x) - Df_t(\phi_t^*)\| = 0,$$

then the following statements hold:

(a) $\max \Sigma_{\mathbb{I}}(\phi^*) < 1$ *if and only if* ϕ^* *is uniformly exponentially stable on* $\mathbb{I}$.

(b) If a spectral interval σ *of* $\Sigma_{\mathbb{I}}(\phi^*)$ *fulfils* $\min \sigma > 1$, *then* ϕ^* *is unstable.*

The dichotomy spectra for autonomous and periodic equations listed in the previous section show that Proposition 5.2.1 generalises the classical stability conditions. Yet, both scalar difference equations $x_{t+1} = x_t + x_t^2$ and $x_{t+1} = x_t - x_t^3$ have the trivial solution with $\Sigma(0) = \{1\}$. Due to [72, p. 29, Theorem 1.5(i)], the zero fixed point of the first equation is unstable, which shows that the converse of statement (b) does not hold. Moreover, [72, p. 29, Theorem 1.15(iii)] implies that the zero solution of the second equation is (uniformly) asymptotically stable, and thus exponential stability in (a) cannot be replaced by asymptotic stability.

A sufficient condition for the exponential stability of ϕ^* in terms of Lyapunov exponents is provided in Theorem A.1.3(b) and requires invertibility of (Y). For related criteria indicating instability, we refer to [154].

We now discuss how stability can be deduced from Lyapunov exponents for a nonlinear skew product flow. Specifically, consider the difference equation

$$x_{t+1} = f(\theta(t,\omega), x_t) \tag{Δ^ω}$$

with a right hand side $f : \Omega \times \mathbb{R}^d \to \mathbb{R}^d$. We assume that the driving system $\theta : \mathbb{Z} \times \Omega \to \Omega$ on the metric space Ω is ergodic with respect to a probability space $(\Omega, \mathfrak{B}(\Omega), \mu)$, and let φ denote the cocycle generated by (Δ^ω). We furthermore assume that for μ-almost all $\omega \in \Omega$, the mapping $x \mapsto f(\omega, x)$ is of class $C^{1,\delta}$, i.e. this function is differentiable and its derivative is Hölder continuous with exponent $\delta \in (0,1)$. We further assume that the (autonomous) skew product system $(\theta, \varphi) : \mathbb{Z}_0^+ \times \Omega \times \mathbb{R}^d \to \mathbb{R}^d$ has an invariant probability measure ρ on $(\Omega \times \mathbb{R}^d, \mathfrak{B}(\Omega) \otimes \mathfrak{B}(\mathbb{R}^d))$, i.e. its marginal on Ω is given by μ, and we have $\rho((\theta, \varphi)^{-1}(1, A)) = \rho(A)$ for all $A \in \mathfrak{B}(\Omega) \otimes \mathfrak{B}(\mathbb{R}^d)$ [7, Definition 1.4.1].

The following linearised stability theorem is taken from [212, Proposition 1] (see also [205, Theorem 5.1] and [206, Theorem 5.1]). It guarantees a type of exponential stability.

Proposition 5.2.2 (Lyapunov Exponents and Linearised Stability) *Under the assumptions*

$$\int_{\Omega \times \mathbb{R}^d} \ln^+ \|D_2 f(\omega, x)\| d\rho(\omega, x) < \infty$$

and

$$\int_{\Omega \times \mathbb{R}^d} \ln^+ \|f(\omega, \cdot + x) - f(\omega, x)\|_{C^{1,\delta}(\bar{B}_1(0))} d\rho(\omega, x) < \infty,$$

the top Lyapunov exponent

$$\lambda(\omega, x) := \limsup_{t \to \infty} \sqrt[t]{\|D_3 \varphi(t, \omega, x)\|}$$

exists for ρ-almost all $(\omega, x) \in \Omega \times \mathbb{R}^d$. Moreover, if there exists some $\gamma < 1$ such that $\lambda(\omega, x) < \gamma$ holds ρ-almost everywhere, then there exist measurable functions $\alpha, \beta : \Omega \times \mathbb{R}^d \to \mathbb{R}^+$ with $\alpha(\omega, x) < \beta(\omega, x)$ for all $(\omega, x) \in \Omega \times \mathbb{R}^d$ such that for ρ-almost all (ω, x), the set

$$S(\omega, x) := \{y \in \bar{B}_{\alpha(\omega,x)}(x) : \|\varphi(t, \omega, y) - \varphi(t, \omega, x)\| \leq \beta(\omega, x)\gamma^t$$

$$\text{for all } t \in \mathbb{Z}_0^+\}$$

is a measurable neighbourhood of x.

5.3 Continuation

This section addresses the continuation of hyperbolic reference solutions of param-eterised equations (Δ_λ). We assume that the difference equation is defined on an unbounded discrete interval of the form $\mathbb{I} = \mathbb{Z}, \mathbb{Z}_0^+, \mathbb{Z}_0^-$ and distinguish between the case of $\mathbb{I} = \mathbb{Z}$ (Sect. 5.3.1) and the half line case (Sect. 5.3.2). The parameter space Λ of (Δ_λ) is given as nonempty open subset of a Banach space Y.

Throughout this section, our assumptions are as follows (see also Hypothe-sis 2.3.1 in the continuous time case).

Hypothesis 5.3.1 *Let* $m \in \mathbb{N}$, *and suppose that each* $f_t : \mathcal{D}(t) \times \Lambda \to \mathbb{R}^d$, $t \in \mathbb{I}$, *is a* C^m-*function such that the following hold for* $0 \leq j \leq m$:

  (i) Uniform boundedness. *For all bounded sets* $B \subseteq \mathcal{D}(t)$, *one has*

$$\sup_{t \in \mathbb{I}} \sup_{x \in B \cap \mathcal{D}(t)} \left| D^j f_t(x, \lambda) \right| < \infty \quad \text{for all } \lambda \in \Lambda.$$

 (ii) Uniform continuity. *For all* $\lambda_0 \in \Lambda$ *and* $\varepsilon > 0$, *there exists a* $\delta > 0$ *with*

$$\sup_{\substack{t \in \mathbb{I} \\ }} \sup_{\substack{x,y \in \mathcal{D}(t) \\ |x-y| < \delta}} \sup_{\lambda \in B_\delta(\lambda_0)} \left| D^j f_t(x, \lambda) - D^j f_t(y, \lambda_0) \right| < \varepsilon.$$

Given a reference solution $\phi^* = (\phi_t^*)_{t \in \mathbb{I}}$ to (Δ_{λ^*}), we denote by $\Sigma_\mathbb{I}(\phi^*)$ the dichotomy spectrum of the variational equation

$$x_{t+1} = D_1 f_t(\phi_t^*, \lambda^*) x_t. \tag{Y_{λ^*}}$$

The question concerning a continuation of the solution ϕ^* under variation of λ near λ^* depends on whether $\mathbb{I}$ is the entire line or a half line.

5.3.1 Continuation of Entire Solutions

The following result says that entire solutions can be uniquely continued provided they are hyperbolic [190, Theorem 2.11].

Theorem 5.3.2 (Continuation of Entire Solutions) *Consider the difference equa-tion* (Δ_λ) *on* $\mathbb{I} = \mathbb{Z}$, *satisfying Hypothesis 5.3.1, and assume that for a given* $\lambda^* \in \Lambda$, *there exists a bounded entire solution* $\phi^* = (\phi_t^*)_{t \in \mathbb{Z}}$ *of* (Δ_{λ^*}) *with*

$$1 \notin \Sigma(\lambda^*).$$

Then there exist $\rho, \varepsilon > 0$ *and a unique* C^m-*function*

$$\phi : B_\rho(\lambda^*) \to B_\varepsilon(\phi^*) \subset \ell^\infty(\mathbb{R}^d)$$

such that $\phi(\lambda^*) = \phi^*$ and every $\phi(\lambda)$ is a bounded and hyperbolic entire solution of (Δ_λ) with the same Morse index as ϕ^*.

Example 1.2.2 illustrates Theorem 5.3.2 in the simple case of a linear difference equation, as well as the situation when hyperbolicity is lost.

Note that Theorem 5.3.2 not only generalises the situation of hyperbolic fixed points for autonomous equations but also ensures that hyperbolic fixed points x^* persist as entire bounded solutions under ℓ^∞-small parametric perturbations, see also Remark 2.3.3 for the continuous time case.

As C^m-function, $\lambda \mapsto \phi(\lambda) \in \ell^\infty(\mathbb{R}^d)$ can be expressed as Taylor series (2.3) (note that we now use the same notation as in Sect. 2.3). The Taylor coefficients

$$D^n \phi(\lambda^*) \in L_n(Y, \ell^\infty(\mathbb{R}^d)) \cong \ell^\infty(L_n(Y, \mathbb{R}^d))$$

fulfil the inhomogeneous linear difference equation

$$X_{t+1} = D_1 f_t(\phi_t^*, \lambda^*) X_t + H_n(t)$$

in $L_n(Y, \mathbb{R}^d)$, where the inhomogeneity $H_n : \mathbb{Z} \to L_n(Y, \mathbb{R}^d)$ reads as

$$H_n(t) y_1 \cdots y_n :=$$
$$\sum_{j=2}^{n} \sum_{(N_1,\ldots,N_j) \in P_j^<(n)} D^j f_t(\phi_t^*, \lambda^*) g_t^{\#N_1}(\lambda^*) y_{N_1} \cdots g_t^{\#N_j}(\lambda^*) y_{N_j}$$

with $g_t^{\#N_1}(\lambda) := \frac{\mathrm{d}^{\#N_1}(\phi(\lambda)_t, \lambda)}{\mathrm{d}\lambda^{\#N_1}}$, and in particular, $H_1(t) = D_2 f_t(\phi_t^*, \lambda^*)$. With these preparations at hand, the following result was deduced in [190, Corollary 2.20].

Proposition 5.3.3 (Taylor Expansions of Continued Entire Solutions) *We consider the situation of Theorem 5.3.2, where the hyperbolic bounded entire solution ϕ^* of (Δ_{λ^*}) was continued via a unique C^m-function $\phi : B_\rho(\lambda^*) \to B_\varepsilon(\phi^*) \subset \ell^\infty(\mathbb{R}^d)$. Then the coefficients $D^n \phi(\lambda^*) : \mathbb{Z} \to L_n(Y, \ell^\infty(\mathbb{R}^d))$ for $n \in \{1, \ldots, m\}$ in the Taylor expansion (2.3) can be determined recursively from the Lyapunov–Perron sums*

$$D^n \phi(\lambda^*) = \sum_{s \in \mathbb{Z}} \Gamma_{\lambda^*}(\cdot, s+1) H_n(s) \quad \text{for all } n \in \{1, \ldots, m\} \, ,$$

where Green's function *for (Y_{λ^*}) is given by*

$$\Gamma_{\lambda^*}(t, s) := \begin{cases} \Phi(t,s) P_s & : \quad s \leq t, \\ -\Phi(t,s)(\mathrm{id} - P_s) & : \quad t < s \end{cases}$$

and $(P_t)_{t \in \mathbb{Z}}$ is the unique invariant projector for the exponential dichotomy of the variational equation (Y_{λ^*}).

Note that in the linear equation (1.11) from Example 1.2.2, one obtains the unique continuation $\phi(\lambda) \in \ell^\infty(\mathbb{R}^d)$ explicitly under the hyperbolicity assumption $|\lambda| \neq 1$.

Example 5.3.4 *With a bounded sequence* $(a_t)_{t \in \mathbb{Z}}$ *in* $\mathbb{R}$, *we consider the parametrically perturbed scalar difference equation*

$$x_{t+1} = \tfrac{4}{\pi} \arctan x_t + \lambda a_t, \qquad (5.3)$$

depending on $\lambda \in \mathbb{R}$. *For* $\lambda = 0$, *the difference equation* (5.3) *is autonomous and admits the three equilibria* $x_0 = 0$ *and* $x_\pm = \pm 1$ *(see Fig. 5.3 (left)). We investigate the behaviour of these fixed points for values* $\lambda \neq 0$.

- $x_0 = 0$: *The variational equation at* $\lambda^* = 0$ *reads as* $x_{t+1} = \tfrac{4}{\pi} x_t$ *and is therefore unstable. Thus, Theorem 5.3.2 ensures that* x_0 *persists for small values of* λ *as an entire bounded solution* $\phi_0(\lambda)$ *to* (5.3). *Due to Proposition 5.3.3, its derivatives can be recursively computed as*

$$D\phi_0(0)_t = -\left(\tfrac{4}{\pi}\right)^{t-1} \sum_{s=t}^{\infty} \left(\tfrac{\pi}{4}\right)^s a_s, \qquad D^2\phi_0(0)_t = 0,$$

$$D^3\phi_0(0)_t = 2\left(\tfrac{4}{\pi}\right)^{t} \sum_{s=t}^{\infty} \left(\tfrac{\pi}{4}\right)^s D\phi_0(0)_s^3, \qquad D^4\phi_0(0)_t = 0 \quad \text{for all } t \in \mathbb{Z}.$$

- $x_\pm = \pm 1$: *Here the variational equation with* $\lambda^* = 0$ *is* $x_{t+1} = \tfrac{2}{\pi} x_t$ *and so* $x_\pm$ *are uniformly asymptotically stable. Their unique continuation* $\phi_\pm(\lambda)$ *for* $\lambda \neq 0$ *can be approximated as above.*

We refer to Fig. 5.3 for the solution portrait of an almost periodically perturbed difference equation (5.3) *for different values of* λ *(see [61, Section 3] for the almost periodicity of* $t \mapsto \sin t$).

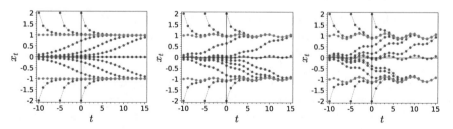

Figure 5.3 Solution sequences (dotted) of the difference equation (5.3) with $a_t = \sin t$ and $\lambda = 0$ (left), $\lambda = 0.05$ (centre) and $\lambda = 0.1$ (right)

5.3.2 *Continuation on Half Lines: Invariant Fibre Bundles*

We next consider the situation where a bounded solution ϕ^* of (Δ_{λ^*}) satisfies a weaker hyperbolicity condition. If we suppose that the variational equation (Y_{λ^*}) has an exponential dichotomy only on a half line $\mathbb{I} = \mathbb{Z}_0^+$ or $\mathbb{I} = \mathbb{Z}_0^-$, then one cannot expect ϕ^* to persist as isolated bounded solution of (Δ_λ) for parameter λ near λ^*. This is illustrated by the following planar system.

Example 5.3.5 *Consider the planar autonomous difference equation*

$$\begin{cases} x_{t+1} = \frac{1}{2}x_t, \\ y_{t+1} = 2y_t + \lambda x_t^2 \end{cases} \tag{5.4}$$

depending on a real parameter λ. From the general solution

$$\varphi_\lambda(t,0,\xi,\eta) = \begin{cases} \begin{pmatrix} \frac{1}{2^t}\xi \\ (\eta + \frac{4}{7}\lambda\xi^2)2^t - \frac{4}{7}\lambda\xi^2\frac{1}{4^t} \end{pmatrix}, & t \geq 0, \\ \begin{pmatrix} \frac{1}{2^t}\xi \\ (\eta + \frac{8}{7}\lambda\xi^2)2^t - \frac{8}{7}\lambda\xi^2\frac{1}{4^t} \end{pmatrix}, & t < 0, \end{cases}$$

we see that the set $W_\lambda^+ := \left\{ (\xi, -\frac{4}{7}\lambda\xi^2) : \xi \in \mathbb{R} \right\}$ consists of all initial values for (5.4) yielding forward bounded solutions, while $W_\lambda^- := \{(0,\eta) : \eta \in \mathbb{R}\}$ captures initial values guaranteeing backward bounded solutions.

From an autonomous perspective, (5.4) possesses a unique hyperbolic fixed point for all $\lambda \in \mathbb{R}$. However, taking a nonautonomous perspective, for $\lambda^* = 0$ the trivial solution $\phi_t^* \equiv 0$ is bounded and hyperbolic on $\mathbb{Z}_0^+$ but embedded into an entire family of such forward bounded hyperbolic solutions, namely those starting in W_0^+. This behaviour even persists for parameters $\lambda \neq \lambda^*$. A dual statement holds for the backward bounded solutions with the set W_λ^-.

Returning to the general case, since the variational equation (Y_{λ^*}) along ϕ^* admits an exponential dichotomy on $\mathbb{I}$, there exists a corresponding so-called stable manifold $\mathcal{W}_{\lambda^*}^+$ (for $\mathbb{I}$ unbounded above), as well as an unstable manifold $\mathcal{W}_{\lambda^*}^-$ (for $\mathbb{I}$ unbounded below) for (Δ_{λ^*}). Hence, a hyperbolic solution ϕ^* on a half line is embedded into a whole family of forward or backward bounded solutions, respectively, making up nonautonomous sets $\mathcal{W}_{\lambda^*}^\pm$. Furthermore, suppose that ϕ^* is embedded into a branch of hyperbolic solutions $\phi(\lambda)$ being smooth in the parameter $\lambda \in \Lambda$. For all $\lambda \in \Lambda$, the *stable nonautonomous set* of $\phi(\lambda)$ is defined as

$$\mathcal{W}_\lambda^+ := \left\{ (\tau,\xi) \in \mathcal{D} : \varphi_\lambda(t,\tau,\xi) - \phi(\lambda)_t \xrightarrow[t\to\infty]{} 0 \right\}$$

and the corresponding *unstable nonautonomous set* reads as

$$\mathcal{W}_\lambda^- := \left\{ (\tau,\xi) \in \mathcal{D} : \begin{array}{l} \text{there exists a solution } \psi = (\psi_t)_{t\in\mathbb{I}} \text{ of } (\Delta_\lambda) \\ \text{such that } \psi_\tau = \xi \text{ and } \psi_t - \phi(\lambda)_t \xrightarrow[t\to-\infty]{} 0 \end{array} \right\},$$

where $\mathbb{I}$ is required to be unbounded above or below, respectively. In the vicinity of $\phi(\lambda)$, these sets can be represented as graphs of a smooth function (this is the *Stable Manifold Theorem*) and one speaks of *invariant fibre bundles* (see Fig. 5.4 for an illustration). These functions depend smoothly on the parameter λ.

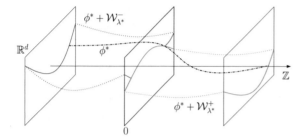

Figure 5.4 Vicinity of a solution ϕ^* (dash dotted, black) having an exponential dichotomy on both half lines $\mathbb{Z}_0^+$ and $\mathbb{Z}_0^-$. Stable bundle $\phi^* + \mathcal{W}_{\lambda^*}^+ \subseteq \mathbb{Z}_0^+ \times \mathbb{R}^d$ (green) and unstable bundle $\phi^* + \mathcal{W}_{\lambda^*}^- \subseteq \mathbb{Z}_0^- \times \mathbb{R}^d$ (red)

Remarks

An early reference for exponential dichotomies in discrete time is [50]. More contemporary approaches give [179] with applications to shadowing. A generalisation to noninvertible difference equations provides [99, pp. 229ff], while [186, pp. 128ff] provides a broad survey of the topic.

Further results can be found in [4] (almost periodic equations allow to conclude an exponential dichotomy on $\mathbb{Z}$ from information over a finite interval). An elegant technique to study dichotomies using operator theory was introduced in [18, 19] (see also [25, 27, 28]). Finally, the dichotomy notion suggested in [15] applies to noninvertible equations but lacks robustness.

Dynamical Spectra For the dichotomy spectrum in discrete time, we refer to [17], while [16] is based on the dichotomy notion from [15] lacking robustness.

The operator-theoretical ideas from [18] were continued in [185]. They resulted in the fine structure of the dichotomy spectrum [192]. Thanks to [180, 224], the scheme of inclusions from Theorem 5.1.8 and the overall spectral theory drastically simplify for almost periodic difference equations [192, Corollary 4.34]. A concept to describe rotation in discrete time might be the *angular values* due to [32, 33].

Multiplicative Ergodic Theorem Deterministic versions of the Multiplicative Ergodic Theorem in discrete time can be found in [7, pp. 135–136, 3.4.2 Proposition] or [53, pp. 231–232, Theorem 11.1.14 and p. 243, Theorem 11.4.1]. In comparison to Theorem 5.1.9, both assume invertibility of the linear cocycle and make stronger statements: they address two-sided time and yield sharp, positive Lyapunov exponents χ_i as identical limits for $t \to \pm\infty$.

As a word of caution: despite the above examples, in general it is difficult to verify an exponential dichotomy rigorously or to compute dichotomy spectra analytically. For a numerical approach to such problems, we refer to [107, 108].

Stability Stability criteria for scalar difference equations in the nonhyperbolic situation $\Sigma_\tau^+(A) = \{1\}$ were given in [194, Proposition 5.4]. This is a special case of the, in general, interesting question for stability conditions in case $1 \in \Sigma_\tau^+(A)$.

Continuation The numerical methods from [111] to compute hyperbolic heteroclinic solutions yield exponential convergence on a smaller centred approximation interval. Thus, this approach also applies in order to compute bounded entire solutions from Theorem 5.3.2 to nonautonomous difference equations, as long as they are hyperbolic.

Chapter 6
Nonautonomous Bifurcation

The reader comparing the section titles of this chapter to the related chapter, Chap. 3, on differential equation might suspect that they are rather similar. However, this impression is deceptive. For instance, Sects. 6.2.1 and 6.3.1 illustrate that rather different phenomena are denoted as a fold bifurcation. As is well known, an attempt to distinguish different nonautonomous bifurcations on a linear level already is given by the fine structure of the dichotomy spectrum from Sect. 5.1. Throughout, at the critical parameter, the dichotomy spectrum contains 1.

The section on attractor bifurcation begins with a review of the time-continuous situation but focuses briefly on the necessary changes; it requires 1 to be contained in the Fredholm spectrum. In contrast, discrete Hopf bifurcations are captured in more detail and by means of a skew product framework.

The situation where 1 is not contained in the Fredholm spectrum allows to introduce several results on solution bifurcation in a rather detailed way, covering fold, transcritical and pitchfork bifurcations. All of them are based on an application of abstract branching theory from for example [132, 229]. The shovel bifurcation finally requires a surjectivity spectrum disjoint from 1, and we merely indicate necessary changes to the rather detailed continuous-time presentation.

When discussing bifurcations of minimal sets and invariant graphs, we restrict to the fold bifurcation. Although it is possibly the simplest bifurcation phenomenon, it nicely illustrates complicated phenomena occurring under quasi-periodic forcing such as the occurrence of strange non-chaotic attractors.

© The Author(s), under exclusive license to Springer Nature Switzerland AG 2023   95
V. Anagnostopoulou et al., *Nonautonomous Bifurcation Theory*, Frontiers
in Applied Dynamical Systems: Reviews and Tutorials 10,
https://doi.org/10.1007/978-3-031-29842-4_6

6.1 Attractor Bifurcation

6.1.1 Transcritical and Pitchfork Bifurcation

The attractor bifurcations from Sect. 3.1 allow analogous counterparts for scalar nonautonomous difference equations (Δ_λ), having a smooth branch $\phi(\lambda)$ of bounded entire solutions. Yet, the corresponding analysis in [198] requires some remarks. We denote the transition matrix of the variational equation

$$x_{t+1} = D_1 f_t(\phi(\lambda)_t, \lambda) x_t \tag{Y_λ}$$

by the real product $\Phi_\lambda(t, s) = \prod_{r=s}^{t-1} D_1 f_r(\phi(\lambda)_r, \lambda)$. The continuous-time Hypothesis 3.1.3 on the transversal exchange of stability needs to be formulated as follows in the discrete-time context.

Hypothesis 6.1.1 (Transversal Exchange of Stability) *We assume invertibility*

$$D_1 f_t(\phi(\lambda)_t, \lambda) > 0 \quad \text{for all } t \in \mathbb{I}', \lambda \in \Lambda \tag{6.1}$$

and that there exist $\lambda^* \in \Lambda$, $K \geq 1$ and functions $\gamma_+, \gamma_- : \Lambda \to (0, \infty)$ either both increasing or both decreasing so that $\lim_{\lambda \to \lambda^*} \gamma_+(\lambda) = \lim_{\lambda \to \lambda^*} \gamma_-(\lambda) = 1$ and

$$\Phi_\lambda(t, s) \leq K(\gamma_+(\lambda))^{t-s}, \quad \Phi_\lambda(s, t) \leq K(\gamma_-(\lambda))^{s-t} \quad \text{for all } t, s \in \mathbb{I}, \lambda \in \Lambda.$$

Although the variational equation needs to be invertible, this does not require global invertibility of $f_t(\cdot, \lambda)$ for $t \in \mathbb{I}'$, something which is not given in various applications. We note that the assumption (6.1) implies at least local invertibility. Under the additional assumption

$$\limsup_{x \to 0} \sup_{\lambda \in \Lambda} \sup_{t \in \mathbb{I}'} [D_1 f_t(x + \phi(\lambda)_t, \lambda) - D_1 f_t(\phi(\lambda)_t, \lambda)] = 0,$$

one obtains a globally invertible extension of the equation of perturbed motion for reference solutions $\phi(\lambda)$ on a neighbourhood of 0 which is uniform in $t \in \mathbb{I}'$ [188, Proposition A.1]. This allows to apply corresponding discrete-time transcritical and pitchfork bifurcation patterns from [198, Theorems 5.1 and 6.1] directly, which give analogous results to Theorems 3.1.4 and 3.1.6.

6.1.2 Sacker–Neimark Bifurcation

The discrete-time analogue to the Hopf bifurcation is given by the Sacker–Neimark bifurcation, and in an autonomous context, this bifurcation involves a complex conjugate pair of eigenvalues [228, pp. 374ff, Section 3.2C]. It is well known that external (random) forcing can lead to the separation of this pair [8], and this gives rise to a two-step bifurcation scenario [7, Chapter 9.4]. The current understanding of the nonautonomous Sacker–Neimark bifurcation is still limited, and in particular, it is an open problem to describe the structure of an invariant torus splitting off. Earlier simulations suggested that this structure is simple, in the sense that the intersection with each fibre of the extended state space is a topological circle [8]. However, later numerical studies indicated that more complicated structures may appear as well [131].

We aim at giving a description of the nonautonomous Sacker–Neimark bifurcation in the context of model systems that are accessible to a rigorous analysis but at the same time allow for highly nontrivial dynamics. These systems exhibit a two-step bifurcation scenario, and additionally, the structure of the split-off torus is simple, in the sense that the intersection with each fibre of the extended state space is a topological circle, but as outlined above, this should not be taken as an indication for the general case.

Throughout the section, let Ω be a compact metric space and $\theta : \mathbb{Z} \times \Omega \to \Omega$ denotes a discrete dynamical system.

Hypothesis 6.1.2 *Suppose that a C^2-function $h : \mathbb{R}_0^+ \to \mathbb{R}_0^+$ is strictly increasing, strictly concave and bounded and satisfies $h(0) = 0$ and $h'(0) = 1$. Moreover, the matrix-valued function $A : \Omega \to SL(\mathbb{R}^2)$ is supposed to be continuous.*

For parameters $\lambda > 0$, consider the planar nonautonomous difference equation

$$\begin{pmatrix} x_{t+1} \\ y_{t+1} \end{pmatrix} = f(\theta(t,\omega), x_t, y_t, \lambda), \tag{6.2}$$

whose right hand side

$$f(\omega, x, y, \lambda) := \frac{h\left(\lambda\sqrt{x^2 + y^2}\right)}{\sqrt{x^2 + y^2}} A(\omega) \begin{pmatrix} x \\ y \end{pmatrix}$$

is continuously extended to $\Omega \times \mathbb{R}^2 \times (0, \infty)$. It gives rise to a parametrised family of skew product flows $(\theta, \varphi_\lambda)$ on $\Omega \times \mathbb{R}^2$.

Such problems were introduced in [93] as examples for the existence of strange non-chaotic attractors (see the remark to Sect. 6.3), while a first step in their rigorous analysis was made in [91], motivating the work in [6].

In order to give a concise description of the bifurcation pattern in (6.2), we concentrate on the behaviour of the global attractor $\mathcal{A}_\lambda$ of $(\theta, \varphi_\lambda)$. Owing to dissipativity properties [139, p. 46, Theorem. 3.20], there exists an $R > 0$ such that the global attractor allows the characterisation

$$\mathcal{A}_\lambda = \bigcap_{t \in \mathbb{N}} (\theta, \varphi_\lambda) \left(\{t\} \times \Omega \times \bar{B}_R(0) \right) \quad \text{for all } \lambda > 0.$$

The subsequent result [6, Theorem 1.1] identifies bifurcations in terms of changes in the global attractor of the system that are induced by changes in the Lyapunov exponents. Thereto, note that the linearisation of (6.2) along the origin is a linear difference equation (Λ^ω) with the Lyapunov exponent $\chi(\omega)$ defined in (5.2). The

critical parameters of the system are, in particular, determined by the maximal Lyapunov exponent of the associated linear cocycle (see discussion after the following
theorem), defined by

$$\chi_{\max}(A) = \sup_{\omega \in \Omega} \limsup_{t \to \infty} \chi(\omega, A) \in (0, 1].$$

Theorem 6.1.3 (A Model for the Nonautonomous Sacker–Neimark Bifurcation)
Consider the family of skew product flows $(\theta, \varphi_\lambda)$ *generated by* (6.2) *and assume
that Hypothesis 6.1.2 holds. If* $\lambda_1^* := \frac{1}{\chi_{\max}(A)}$ *and* $\lambda_2^* := \chi_{\max}(A)$, *then the following supercritical bifurcation holds:*

(a) *If* $\lambda < \lambda_1^*$, *then the global attractor* $\mathcal{A}_\lambda$ *is* $\Omega \times \{(0,0)\}$.

(b) *If* $\lambda_1^* < \lambda < \lambda_2^*$, *then there exists at least one* $\omega \in \Omega$ *such that* $\mathcal{A}_\lambda(\omega)$ *is a
line segment of positive length.*

(c) *If* $\lambda > \lambda_2^*$, *then for all* $\omega \in \Omega$, *the fibre* $\mathcal{A}_\lambda(\omega)$ *is a closed topological disc
that depends continuously on* ω. *Furthermore, the compact* $(\theta, \varphi_\lambda)$-*invariant
set* $\mathcal{T}_\lambda = \partial \mathcal{A}_\lambda$ *is the global attractor outside* $\Omega \times \{(0,0)\}$ *in the sense that*

$$\mathcal{T}_\lambda = \bigcap_{t \in \mathbb{N}} (\theta, \varphi_\lambda) \left(\{t\} \times \Omega \times \left(\bar{B}_1(0) \setminus \bar{B}_\delta(0) \right) \right)$$

for all sufficiently small $\delta > 0$.

An illustration of Theorem 6.1.3 was given in Fig. 1.9.

The most important property of equation (6.2) is the fact that its fibre maps send
lines passing through the origin to such lines again. As a consequence, the right hand
side written in polar coordinates gives rise to a double skew product, which is crucial
for the analysis of the system, since it is now possible to apply existing results on
skew product maps with one-dimensional fibres to study the dynamics of the polar
coordinate system (see [6, Section 3.2]); we illustrate this in Example 6.1.5. Yet,
the fact that $A(\cdot)$ can be chosen arbitrarily allows for a rich variety of dynamics
inside the global attractor when $\lambda > \lambda_2^*$, such as minimal dynamics, the existence
of attractor–repeller pairs or more complicated behaviour [6].

Remark 6.1.4 (Intermediate Region) *For parameters* $\lambda \in (\lambda_1^*, \lambda_2^*)$, *as well as in
the critical cases* $\lambda = \lambda_1^*$ *and* $\lambda = \lambda_2^*$, *the dynamical behaviour is more intricate
and in particular it is not uniform for all orbits; given two* θ-*invariant measures* μ_1
and μ_2 *on* Ω, *the typical dynamics with respect to* μ_1 *and* μ_2 *may be very different.
To take this into account, an approach using random dynamical systems is helpful:
one can derive a random analogue [6, Theorem 1.3] of Theorem 6.1.3, where the
topological structure of* Ω *is replaced by a measure-theoretical one. Consequently,
$\mathcal{A}_\lambda$ *has no global topological structure either and the focus is on the structure of
$\mathcal{A}_\lambda$ *on typical fibres. The critical parameters (which are now* μ-*dependent) are determined by the top Lyapunov exponent of the linear skew product flow* (θ, A) *with
respect to a measure* μ, *which reads as*

$$\chi_\mu(A) = \lim_{t\to\infty} \int_\Omega \sqrt[t]{\|\Phi(t,\omega)\|} \, d\mu(\omega).$$

In the intermediate region $\frac{1}{\chi_\mu(A)} < \lambda < \chi_\mu(A)$*, there exists a set* $\Omega_0 \subseteq \Omega$ *of full measure, such that the set* $A_\lambda(\omega)$ *is a line segment of positive length for all* $\omega \in \Omega_0$. *Moreover, there is a random two-point forward attractor embedded within* $A_\lambda(\omega)$, *which consists exactly of the endpoints of the line segment* $A_\lambda(\omega)$ *on each fibre.*

If we additionally have the topological setting of Theorem 6.1.3, and $M(\theta)$ denotes the set of θ-invariant ergodic probability measures on Ω, then a straightforward consequence of the semi-uniform sub-multiplicative ergodic theorem [222, Corollary 1.11] yields that $\chi_{\max}(A) = \sup_{\mu \in M(\theta)} \chi_\mu(A)$. Due to compactness of the space $M(\theta)$, there always exists at least one $\hat{\mu} \in M(\theta)$ that achieves this supremum. Thus, for $\lambda \in (\lambda_1^*, \lambda_2^*)$, this means, in particular, that $\hat{\mu}$-typical fibres are line segments of positive length and the typical dynamics with respect to $\hat{\mu}$ are governed by the aforementioned two-point attractor; Theorem 6.1.3(b) is a direct consequence of this.

Our Example 1.2.8 illustrates Theorem 6.1.3 with quasi-periodic forcing θ. Due to the unique ergodicity of θ, in the intermediate region, all fibres of the global attractor are line segments of positive length, as shown in Fig. 6.1a. The following example continues Example 1.2.8 and illustrates the bifurcation of invariant graphs

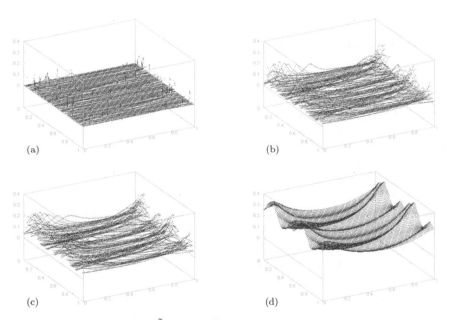

(a)

(b)

(c)

(d)

Figure 6.1 The global attractor $\tilde{A}_\lambda$ of the induced polar coordinate system from Example 6.1.5. (a) and (b) show the $\tilde{A}_\lambda$ shortly after the first critical parameter λ_1^* and just before the second λ_2^*, respectively. (c) shows $\tilde{A}_\lambda$ shortly after λ_1^* where the invariant torus has just formed, and finally, (d) depicts the split-off torus far from the bifurcation

taking place in the polar coordinate system (for more on bifurcations of invariant graphs, see Sect. 6.3).

Example 6.1.5 *In Example 1.2.8, we consider a nonautonomous difference equation (1.18), which corresponds to (6.2) with the function $h(x) := \frac{1}{3\sqrt{2}} \arctan x$ and*

$$A(\omega) = \begin{pmatrix} \frac{1}{\sqrt{2}} & 0 \\ 0 & \sqrt{2} \end{pmatrix} \begin{pmatrix} \cos(2\pi\omega) & \sin(2\pi\omega) \\ -\sin(2\pi\omega) & \cos(2\pi\omega) \end{pmatrix} =: \begin{pmatrix} a_{11}(\omega) & a_{12}(\omega) \\ a_{21}(\omega) & a_{22}(\omega) \end{pmatrix}.$$

In Euclidean coordinates $(x, y) \in \mathbb{R}^2$, it generates a skew product flow $(\theta, \varphi_\lambda)$ on $\Omega \times \mathbb{R}^2$. In polar coordinates $(\vartheta, \rho) \in (\mathbb{R}/\mathbb{Z}) \times (0, \infty)$, however, a simple projective coordinate transformation yields another skew product flow (Θ, ρ_λ). Its driving system $\Theta : \mathbb{Z}_0^+ \times \Omega \times \mathbb{R}/\mathbb{Z} \to \Omega \times \mathbb{R}/\mathbb{Z}$ (here only defined in forward time), in turn, is a skew product

$$\Theta(t, \omega, \alpha) = \begin{pmatrix} \theta(t, \omega) \\ \vartheta(t, \omega, \alpha) \end{pmatrix}$$

with $\vartheta(1, \omega, \alpha) := \frac{1}{\pi} \arctan \frac{a_{21}(\omega) + a_{22}(\omega) \tan(\pi\alpha)}{a_{11}(\omega) + a_{12}(\omega) \tan(\pi\alpha)}$ (mod 1) and furthermore the cocyle $\rho_\lambda : \mathbb{Z}_0^+ \times \Omega \times \mathbb{R}/\mathbb{Z} \times [0, 1] \to [0, 1]$,

$$\rho_\lambda(1, \omega, \alpha, r) := \frac{\arctan(\lambda r)}{3\sqrt{2}} \left\| \Phi(t, \omega) \begin{pmatrix} \cos(\pi\alpha) \\ \sin(\pi\alpha) \end{pmatrix} \right\|.$$

The global attractor of (Θ, ρ_λ) is given by

$$\tilde{A}_\lambda = \bigcap_{t \geq 0} (\Theta, \rho_\lambda) \left(\{t\} \times \Omega \times \mathbb{R}/\mathbb{Z} \times [0, 1] \right).$$

Due to the monotonicity of ρ_λ in the fourth argument r, we can define an invariant graph $\psi_\lambda^+ : \Omega \times \mathbb{R}/\mathbb{Z} \to [0, 1]$ as the upper bounding graph *of $\tilde{A}_\lambda$, given by*

$$\psi_\lambda^+(\omega, \alpha) = \sup \tilde{A}_\lambda(\omega, \alpha) \quad \text{for all } (\alpha, \omega) \in \Omega \times \mathbb{R}/\mathbb{Z}.$$

Note that $\tilde{A}_\lambda = \{(\omega, \alpha, r) \in \Omega \times \mathbb{R}/\mathbb{Z} \times \mathbb{R}_0^+ : r \in [0, \psi_\lambda^+(\omega, \alpha)]\}$, and a second invariant graph $\psi^-(\omega, \alpha) \equiv 0$ exists (which may be equal to ψ_λ^+). Then, depending on the bifurcation parameter λ, we obtain the following behaviour:

(i) *If $\lambda < \lambda_1^*$, then the global attractor for the polar coordinate system is equal to $\Omega \times \mathbb{R}/\mathbb{Z} \times \{0\}$. In particular, ψ^- is the unique invariant graph of the system, and $\lim_{t \to \infty} \rho_\lambda(t, \omega, \alpha, r) = 0$ for all $(\omega, \alpha, r) \in \Omega \times \mathbb{R}/\mathbb{Z} \times [0, \infty)$.*

(ii) *If $\lambda_1^* < \lambda < \lambda_2^*$, then for all $\omega \in \Omega$, there exists a function $\eta_u(\omega) : \Omega \to \mathbb{R}/\mathbb{Z}$ such that $\psi_\lambda^+(\omega, \eta_u(\omega)) > 0$ and $\psi_\lambda^+(\omega, \alpha) = 0$ for all $\alpha \in (\mathbb{R}/\mathbb{Z}) \setminus \{\eta_u(\omega)\}$.*

(iii) *If $\lambda > \lambda_2^*$, then the invariant graph $\psi_\lambda^+ : \Omega \times \mathbb{R}/\mathbb{Z} \to [0, 1]$ is continuous and strictly positive. It is attracting in the sense that*

$$\lim_{t\to\infty} \left(\rho_\lambda(t,\omega,\alpha,r) - \psi_\lambda^+(\Theta(t,\omega,\alpha)) \right) = 0$$

*for all $(\omega,\alpha,r) \in \Omega \times \mathbb{R}/\mathbb{Z} \times (0,\infty)$. The invariant graph ψ_λ^- is repelling.
If also $\Omega = \mathbb{R}/\mathbb{Z}$ and $\theta : \mathbb{Z} \times \Omega \to \Omega$ describes the irrational rotation defined
in (1.16), then recall from Example 1.2.8 that $\lambda_1^* = 4$ and $\lambda_2^* = \frac{9}{2}$. In addition,
the behaviour of the global attractor $\tilde{A}_\lambda$ and the behaviour of the invariant graphs
$\psi_\lambda^-, \psi_\lambda^+$ of the induced polar coordinate system are illustrated in Fig. 6.1.*

6.2 Solution Bifurcation

In this section, we present several types of solution bifurcations where an exponential dichotomy on the entire line breaks down in the sense that Theorem 5.1.3(b) does not hold. Here the existence of exponential dichotomies is guaranteed on both half lines, and this enables the use of an abstract analytical branching theory based on Fredholm linearisations. Rather than using dynamical systems tools, we consider difference equations as abstract equations in sequence spaces. In this sense, our approach is consistent with the strategy in Sect. 5.3. Of particular importance in this functional analytical approach will be solutions in the space of bounded sequences $\ell^\infty(\mathbb{R}^d)$.

Based on the same definition of a *solution bifurcation* as in continuous time given in Sect. 3.2, a necessary condition follows readily, which can be proved as in Proposition 3.2.1, using Theorem 5.3.2.

Proposition 6.2.1 (Necessary Condition for Bifurcation) *Let $\lambda^* \in \Lambda$. If a
bounded entire solution ϕ^* of (Δ_{λ^*}) bifurcates at λ^*, then ϕ^* is not hyperbolic
on $\mathbb{Z}$.*

In order to ensure nonhyperbolicity, the characterisation given in Theorem 5.1.3 offers two possibilities: on the one hand, for instance assumptions like in Hypothesis 6.1.1 violate the condition Theorem 5.1.3(a). On the other hand, we now violate Theorem 5.1.3(b) and make the following crucial and standing assumption.

Hypothesis 6.2.2 *Let $\lambda^* \in \Lambda$ and ϕ^* be a bounded entire solution of (Δ_{λ^*}). Suppose that the variational equation (Y_{λ^*}) has an exponential dichotomy on both $\mathbb{Z}_0^+$
(with projector P_t^+) and on $\mathbb{Z}_0^-$ (with projector P_t^-) such that there exist nonzero
$v,w \in \mathbb{R}^d$ satisfying*

$$R(P_0^+) \cap N(P_0^-) = \mathbb{R}v \quad and \quad (R(P_0^+) + N(P_0^-))^\perp = \mathbb{R}w. \tag{6.3}$$

We note that Hypothesis 6.2.2 cannot hold with trivial projectors $P_0^+ \in \{0, \mathrm{id}\}$ and for one-dimensional state spaces, i.e. $d = 1$. In terms of the fine structure, it implies that $1 \in \Sigma_p(\lambda^*)$, and we have exponential dichotomies on both half lines and thus $1 \in \Sigma(\lambda^*) \setminus \Sigma_{F_0}(\lambda^*)$ [192, Proposition 4.9].

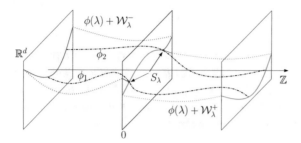

Figure 6.2 Intersection $S_\lambda \subset \mathbb{R}^d$ of the stable bundle $\phi(\lambda) + \mathcal{W}_\lambda^+ \subseteq \mathbb{Z}_0^+ \times \mathbb{R}^d$ (green) with the unstable bundle $\phi(\lambda) + \mathcal{W}_\lambda^- \subseteq \mathbb{Z}_0^- \times \mathbb{R}^d$ (red) at time $t = 0$ yields two bounded entire solutions ϕ_1, ϕ_2 (dash dotted, black) to a difference equation (Δ_λ) indicated as dotted dashed lines

Most importantly, the bifurcation scenarios in the upcoming subsections correspond to Hypothesis 6.2.2 and are thus exclusively observed in nonautonomous systems, in the sense that autonomous, periodic and even almost periodic equations do not satisfy Hypothesis 6.2.2. Indeed, if the variational equation (Y_{λ^*}) is almost periodic, then an exponential dichotomy on a half line extends to the entire integer line [224, Theorem 2], and the reference solution ϕ^* becomes hyperbolic.

From an abstract functional analytical perspective, Hypothesis 6.2.2 guarantees Fredholm properties of the linearisation and thus allows to employ a Lyapunov–Schmidt reduction technique [132, 229]. But Hypothesis 6.2.2 also yields a geometrical insight into the following bifurcation results using *invariant fibre bundles*, i.e. nonautonomous counterparts to invariant manifolds: since the variational equation (Y_{λ^*}) has an exponential dichotomy on $\mathbb{Z}_0^+$, there exists a stable fibre bundle $\phi^* + \mathcal{W}_{\lambda^*}^+$ consisting of all solutions to (Δ_{λ^*}) approaching ϕ^* in forward time. In particular, $\mathcal{W}_{\lambda^*}^+$ is locally a graph over the stable nonautonomous set $\mathcal{V}^+$ of the linearisation. Analogously, an exponential dichotomy on the negative half line $\mathbb{Z}_0^-$ guarantees an unstable fibre bundle $\phi^* + \mathcal{W}_{\lambda^*}^-$ consisting of solutions decaying to ϕ^* in backward time. If ϕ^* is embedded in a smooth branch $\phi(\lambda)$ of bounded entire solutions as described in Sect. 5.3.2, then the fibre bundles persist for λ near λ^* as graphs $\mathcal{W}_\lambda^+, \mathcal{W}_\lambda^-$. Hence, bounded entire solutions to (Δ_{λ^*}) are contained in the set $(\phi^* + \mathcal{W}_{\lambda^*}^+) \cap (\phi^* + \mathcal{W}_{\lambda^*}^-)$. For λ near λ^*, one concludes that the intersection

$$S_\lambda := \phi(\lambda)_0 + \mathcal{W}_\lambda^+(0) \cap \phi(\lambda)_0 + \mathcal{W}_\lambda^-(0) \subset \mathbb{R}^d$$

yields initial values (at time 0) for bounded entire solutions to (Δ_λ) (see Fig. 6.2).

6.2.1 Fold Bifurcation

We first study a fold bifurcation scenario from [187, Theorem 2.13].

Theorem 6.2.3 (Fold Solution Bifurcation) *Suppose that Hypothesis 6.2.2 holds and that (Δ_λ) is of class C^m, where $m \geq 2$. If*

$$g_{01} := \sum_{s \in \mathbb{Z}} \langle \Phi_{\lambda^*}(0, s+1)^T w, D_2 f_s(\phi_s^*, \lambda^*) \rangle \neq 0 ,$$

then there exist a real $\rho > 0$, open convex neighbourhoods $U \subseteq \ell^\infty(\mathbb{R}^d)$ of ϕ^* and $\Lambda_0 \subseteq \Lambda$ of λ^*, and C^m-functions $\phi : (-\rho, \rho) \to U$, $\lambda : (-\rho, \rho) \to \Lambda_0$ so that
  (a) $\phi(0) = \phi^*$, $\lambda(0) = \lambda^*$ and $\dot\phi(0) = \Phi_{\lambda^*}(\cdot, 0)v$, $\dot\lambda(0) = 0$.
  (b) *Each $\phi(s)$ is an entire solution of $(\Delta_{\lambda(s)})$ in $\ell^\infty(\mathbb{R}^d)$.*
Under the additional assumption

$$g_{20} := \sum_{s \in \mathbb{Z}} \langle \Phi_{\lambda^*}(0, s+1)^T w, D_1^2 f_s(\phi_s^*, \lambda^*)(\Phi_{\lambda^*}(s, 0)v)^2 \rangle \neq 0 ,$$

the solution $\phi^* \in \ell^\infty(\mathbb{R}^d)$ of (Δ_{λ^*}) bifurcates at λ^*, one has $\ddot\lambda(0) = -\frac{g_{20}}{g_{01}}$ and the following holds locally in $U \times \Lambda_0$.
  (c) *Subcritical case. If $g_{20}/g_{01} > 0$, then (Δ_λ) has no entire solution in $\ell^\infty(\mathbb{R}^d)$ for $\lambda > \lambda^*$, ϕ^* is the unique entire solution of (Δ_{λ^*}) in $\ell^\infty(\mathbb{R}^d)$ and (Δ_λ) has exactly two distinct entire bounded solutions for $\lambda < \lambda^*$.*
  (d) *Supercritical case. If $g_{20}/g_{01} < 0$, then (Δ_λ) has no entire solution in $\ell^\infty(\mathbb{R}^d)$ for $\lambda < \lambda^*$, ϕ^* is the unique entire solution of (Δ_{λ^*}) in $\ell^\infty(\mathbb{R}^d)$ and (Δ_λ) has exactly two distinct entire bounded solutions for $\lambda > \lambda^*$.*

We illustrate this theorem by means of the following example.

Example 6.2.4 *Consider the planar equation*

$$\begin{pmatrix} x_{t+1} \\ y_{t+1} \end{pmatrix} = \begin{pmatrix} b_t & 0 \\ 0 & c_t \end{pmatrix} \begin{pmatrix} x_t \\ y_t \end{pmatrix} + \begin{pmatrix} 0 \\ x_t^2 - \lambda \end{pmatrix} =: f_t(x_t, y_t, \lambda) , \tag{6.4}$$

where

$$b_t := \begin{cases} 2 & : \ t < 0, \\ \frac{1}{2} & : \ t \geq 0, \end{cases} \qquad c_t := \begin{cases} \frac{1}{2} & : \ t < 0, \\ 2 & : \ t \geq 0 \end{cases} \tag{6.5}$$

are asymptotically constant sequences. Let $\varphi_\lambda(\cdot, 0, \xi, \eta)$ denote the process generated by (6.4). Its first component φ_λ^1 reads as

$$\varphi_\lambda^1(t, 0, \xi, \eta) = 2^{-|t|}\xi \quad \text{for all } t \in \mathbb{Z}, \tag{6.6}$$

while the Variation of Constants formula yields the asymptotic representation

$$\varphi_\lambda^2(t, 0, \xi, \eta) = \begin{cases} 2^t \left(\eta + \frac{4}{7}\xi^2 - \lambda \right) + O(1) & \text{as } t \to \infty, \\ \frac{1}{2^t} \left(\eta - \frac{1}{2}\xi^2 + 2\lambda \right) + O(1) & \text{as } t \to -\infty. \end{cases}$$

This means that the sequence $\varphi_\lambda(\cdot, 0, \xi, \eta)$ is bounded if and only if $\eta = -\frac{4}{7}\xi^2 + \lambda$ and $\eta = \frac{1}{2}\xi^2 - 2\lambda$ holds, i.e. $\xi^2 = \frac{7}{2}\lambda$ and $\eta = -\lambda$. It follows that there exist two bounded solutions if $\lambda > 0$, and the trivial solution is the unique bounded solution for $\lambda = 0$. There are no bounded solutions for parameters $\lambda < 0$, see Fig. 6.3 for an illustration. Hence, $(\phi^*, \lambda^*) = (0, 0)$ is a bifurcation point, since the number of bounded entire solutions increases from 0 to 2.

This can be deduced from Theorem 6.2.3 as well. The variational equation for (6.4) corresponding to the zero solution and the critical parameter $\lambda^* = 0$ reads as

$$\begin{pmatrix} x_{t+1} \\ y_{t+1} \end{pmatrix} = D_1 f_t(0, 0, 0) \begin{pmatrix} x_t \\ y_t \end{pmatrix} = \begin{pmatrix} b_t & 0 \\ 0 & c_t \end{pmatrix} \begin{pmatrix} x_t \\ y_t \end{pmatrix}.$$

It admits an exponential dichotomy on $\mathbb{Z}_0^+$, as well as on $\mathbb{Z}_0^-$ with respective invariant projectors $P_t^+ \equiv \begin{pmatrix} 1 & 0 \\ 0 & 0 \end{pmatrix}$ and $P_t^- \equiv \begin{pmatrix} 0 & 0 \\ 0 & 1 \end{pmatrix}$. This yields

$$R(P_0^+) \cap N(P_0^-) = \mathbb{R} \begin{pmatrix} 1 \\ 0 \end{pmatrix} \quad \text{and} \quad R(P_0^+) + N(P_0^-) = \mathbb{R} \begin{pmatrix} 1 \\ 0 \end{pmatrix},$$

and (b) of Theorem 5.1.3 is violated. Hence, the trivial solution to (6.4) is not hyperbolic for $\lambda = 0$. On the other hand, Hypothesis 6.2.2 holds with $v = \begin{pmatrix} 1 \\ 0 \end{pmatrix}$ and $w = \begin{pmatrix} 0 \\ 1 \end{pmatrix}$. We compute

$$g_{01} = -\sum_{s \in \mathbb{Z}} \left(\tfrac{1}{2}\right)^{|s+1|} = -3 \quad \text{and} \quad g_{20} = \tfrac{12}{7},$$

and Theorem 6.2.3 yields that the bounded solutions to (6.4) exhibit a supercritical fold bifurcation. This corresponds to the explicit results from Example 6.2.4.

A similar analysis can also be applied to the related equation

$$\begin{pmatrix} x_{t+1} \\ y_{t+1} \end{pmatrix} = \begin{pmatrix} b_t & 0 \\ 0 & c_t \end{pmatrix} \begin{pmatrix} x_t \\ y_t \end{pmatrix} + \begin{pmatrix} 0 \\ x_t^3 - \lambda \end{pmatrix} \tag{6.7}$$

with a cubic, rather than a quadratic nonlinearity as previously in (6.4). Using the Variation of Constants formula, one derives that the crucial second component for the general solution $\varphi_\lambda(\cdot, 0, \xi, \eta)$ to (6.7) fulfils

$$\varphi_\lambda^2(t, 0, \xi, \eta) = \begin{cases} 2^t \left(\eta + \frac{8}{15}\xi^3 - \lambda\right) + O(1) & \text{as } t \to \infty, \\ \frac{1}{2^t}\left(\eta - \frac{2}{15}\xi^3 + 2\lambda\right) + O(1) & \text{as } t \to -\infty. \end{cases}$$

Since its first component is also given by (6.6), the sequence $\varphi_\lambda(\cdot, 0, \xi, \eta)$ is bounded if and only if $\eta = -\frac{8}{15}\xi^3 + \lambda$ and $\eta = \frac{2}{15}\xi^3 - 2\lambda$, which is equivalent to $\xi = \sqrt[3]{4.5\lambda}$ and $\eta = -\frac{7}{5}\lambda$. Thus, initial values $(\xi, \eta) \in \mathbb{R}^2$ on the cusp shaped curve in Fig. 6.4 lead to bounded entire solutions of (6.7) and there is no bifurcation.

We note that a rather different fold bifurcation scenario in the context of skew product flows is described in Sect. 6.3.

Figure 6.3 Supercritical fold
solution bifurcation from
Example 6.2.4 with $\lambda^* = 0$
and different parameters λ:
initial values $(\xi, \eta) \in \mathbb{R}^2$
guaranteeing the existence of
an entire bounded solution
$\varphi_\lambda(\cdot, 0, \xi, \eta)$ of (6.4)

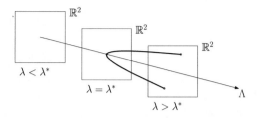

Figure 6.4 Cusp in Example 6.2.4 with $\lambda^* = 0$ and
different parameters λ: initial
values $(\xi, \eta) \in \mathbb{R}^2$ yielding
an entire bounded solution
$\varphi_\lambda(\cdot, 0, \xi, \eta)$ of (6.7)

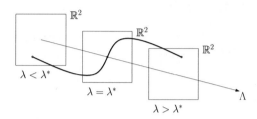

6.2.2 Crossing Curve Bifurcation

Further prototype bifurcation patterns for equations possessing a trivial branch of
solutions are of transcritical and pitchfork type. In this context, it is clear that any
branch $\phi(\lambda)$ of solutions to (Δ_λ) can be transformed into the trivial one, as long
as $\phi(\lambda)$ is known beforehand. The following result does not require such global
information and contains pitchfork and transcritical bifurcations as special cases.

Theorem 6.2.5 (Crossing Curve Solution Bifurcation) *We suppose that Hypoth-
esis 6.2.2 holds and that* (Δ_λ) *is of class* C^m, *where* $m \geq 2$. *If* $D_2 f_t(\phi_t^*, \lambda^*) = 0$
for all $t \in \mathbb{Z}$,

$$g_{02} := \sum_{s \in \mathbb{Z}} \langle \Phi_{\lambda^*}(0, s+1)^T w, D_2^2 f_s(\phi_s^*, \lambda^*) \rangle = 0, \qquad (6.8)$$

and the transversality condition

$$g_{11} := \sum_{s \in \mathbb{Z}} \langle \Phi_{\lambda^*}(0, s+1)^T w, D_1 D_2 f_s(\phi_s^*, \lambda^*) \Phi_{\lambda^*}(s, 0) v \rangle \neq 0$$

holds and then the entire solution ϕ^* *of* (Δ_{λ^*}) *bifurcates at* λ^*. *In detail, there exist
open convex neighbourhoods* $(-\rho, \rho) \subseteq \mathbb{R}$ *of* 0, $U_1 \times U_2 \subseteq \ell^\infty(\mathbb{R}^d) \times \Lambda$ *of* (ϕ^*, λ^*)
and C^{m-1}-*curves* $\gamma_1, \gamma_2 : (-\rho, \rho) \to U_1 \times U_2$ *with the following properties:*

(a) *The set of bounded entire solutions for* (Δ_λ) *in* U_1 *is given by the intersection*
   $(\gamma_1((-\rho, \rho)) \cup \gamma_2((-\rho, \rho))) \cap \ell^\infty(\mathbb{R}^d) \times \{\lambda\}$, *see Fig. 6.5.*

(b) $\gamma_1(u) = (\gamma(u), \lambda^* + u)$ *with a* C^{m-1}-*function* $\gamma : (-\rho, \rho) \to U_1$ *and moreover* $\gamma_1(0) = (\phi^*, \lambda^*)$, $\dot{\gamma}(0) = 0$,

$$\gamma_2(0) = \begin{pmatrix} \phi^* \\ \lambda^* \end{pmatrix} \quad \text{and} \quad \dot{\gamma}_2(0) = \begin{pmatrix} \Phi_{\lambda^*}(\cdot,0)v \\ -\frac{g_{20}}{2g_{11}} \end{pmatrix},$$

where $g_{20} := \sum_{s \in \mathbb{Z}} \langle \Phi_{\lambda^*}(0, s+1)^T w, D_1^2 f_s(\phi_s^*, \lambda^*)(\Phi_{\lambda^*}(s,0)v)^2 \rangle.$

We note that this result is a discrete-time counterpart to [193, Theorem 4.1] and can be proved analogously. If the reference solution ϕ^* is embedded into a branch of trivial solutions to (Δ_λ), then (6.8) is automatically fulfilled and γ_1, respectively, γ, represents the zero branch. In this sense, Theorem 6.2.5 generalises [187, Theorem 3.14 and Corollaries 3.15, 3.16].

The following corollary sheds some light on the structure of the crossing curve solution bifurcation pattern. This requires introducing *Green's function*

$$\Gamma_P(t,s) := \begin{cases} \Phi_{\lambda^*}(t,s)P_s & : \quad t \geq s, \\ -\Phi_{\lambda^*}(t,s)(\mathrm{id} - P_s) & : \quad t < s, \end{cases}$$

with one of the projectors $P \in \{P^-, P^+\}$, and for a given sequence $\psi = (\psi_t)_{t \in \mathbb{Z}}$, we use the notation

$$\overline{\psi_s} := \begin{cases} \Phi_{\lambda^*}(s,0)P_0^+ \xi_0^* + \sum_{r=0}^{\infty} \Gamma_{P^+}(s, r+1)\psi_r & : \quad s \geq 0, \\ \Phi_{\lambda^*}(s,0)[\mathrm{id} - P_0^-]\xi_0^* + \sum_{r=-\infty}^{-1} \Gamma_{P^-}(s, r+1)\psi_r & : \quad s < 0, \end{cases} \tag{6.9}$$

$$\xi_0^* := [P_0^+ + P_0^- - \mathrm{id}]^\dagger \left(\sum_{r=-\infty}^{-1} \Phi_{\lambda^*}(0, r+1)P_r^- \psi_r \right.$$
$$\left. + \sum_{r=0}^{\infty} \Phi_{\lambda^*}(0, r+1)(\mathrm{id} - P_r^+)\psi_r \right),$$

and $[P_0^+ + P_0^- - \mathrm{id}]^\dagger \in \mathbb{R}^{d \times d}$ denotes the pseudo-inverse to $P_0^+ + P_0^- - \mathrm{id}$, see [41].

Corollary 6.2.6 (Transcritical and Pitchfork Solution Bifurcation) *Assume that Hypothesis 6.2.2 holds and that (Δ_λ) is of class C^m, where $m \geq 2$. Then the following statements hold:*

(a) *For $g_{20} \neq 0$, there are locally exactly two entire solutions to (Δ_λ) in $\ell^\infty(\mathbb{R}^d)$ for $\lambda \neq \lambda^*$. This yields a transcritical bifurcation, see Fig. 6.5 (left).*

(b) *In the degenerate case $g_{20} = 0$, we assume that $m \geq 3$ and a higher order condition (with the overlined expression defined according to (6.9))*

$$g_{30} := \sum_{s \in \mathbb{Z}} \langle \Phi_{\lambda^*}(0, s+1)^T w, D_1^3 f_s(\phi_s^*, \lambda^*)(\Phi_{\lambda^*}(s,0)v)^3 \rangle$$
$$- 3 \sum_{s \in \mathbb{Z}} \langle \Phi_{\lambda^*}(0, s+1)^T w,$$
$$D_1^2 f_s(\phi_s^*, \lambda^*)\Phi_{\lambda^*}(s,0)v \overline{D_1^2 f_s(\phi_s^*, \lambda^*)(\Phi_{\lambda^*}(s,0)v)^2} \rangle \neq 0,$$

yields a pitchfork bifurcation, see Fig. 6.5 (right).

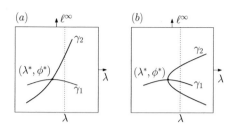

Figure 6.5 Schematic crossing curve bifurcation from Theorem 6.2.5:
(**a**) generic case of a transcritical bifurcation and
(**b**) degenerate case of a supercritical pitchfork bifurcation

(c) Supercritical case. For $g_{30}/g_{11} < 0$, locally there is a unique entire solution of (Δ_λ) in $\ell^\infty(\mathbb{R}^d)$ for $\lambda \leq \lambda^*$, and (Δ_λ) has exactly three entire solutions in $\ell^\infty(\mathbb{R}^d)$ for $\lambda > \lambda^*$.

(d) Subcritical case. For $g_{30}/g_{11} > 0$, locally there is a unique entire solution of (Δ_λ) in $\ell^\infty(\mathbb{R}^d)$ for $\lambda \geq \lambda^*$, and (Δ_λ) has exactly three entire solutions in $\ell^\infty(\mathbb{R}^d)$ for $\lambda < \lambda^*$.

The following two examples illustrate Corollary 6.2.6.

Example 6.2.7 (Transcritical Solution Bifurcation) *Consider the nonlinear difference equation*

$$\begin{pmatrix} x_{t+1} \\ y_{t+1} \end{pmatrix} = f_t(x_t, y_t, \lambda) := \begin{pmatrix} b_t & 0 \\ \lambda & c_t \end{pmatrix} \begin{pmatrix} x_t \\ y_t \end{pmatrix} + \begin{pmatrix} 0 \\ x_t^2 \end{pmatrix}$$

with sequences $(b_t)_{t\in\mathbb{Z}}$ and $(c_t)_{t\in\mathbb{Z}}$ as defined in (6.5). As seen in Example 6.2.4, our assumptions hold with $\lambda^* = 0$ and $g_{11} = \frac{4}{3} \neq 0$, $g_{20} = \frac{12}{7} \neq 0$. Hence, Corollary 6.2.6(a) can be applied and yields that the trivial solution has a transcritical bifurcation at $\lambda = 0$. This was described quantitatively in Example 1.2.9 and illustrated in Fig. 1.10.

Example 6.2.8 (Pitchfork Solution Bifurcation) *Let $\delta \neq 0$ be a fixed real number, and consider the nonlinear difference equation*

$$\begin{pmatrix} x_{t+1} \\ y_{t+1} \end{pmatrix} = f_t(x_t, y_t, \lambda) := \begin{pmatrix} b_t & 0 \\ \lambda & c_t \end{pmatrix} \begin{pmatrix} x_t \\ y_t \end{pmatrix} + \delta \begin{pmatrix} 0 \\ x_t^3 \end{pmatrix} \tag{6.10}$$

with sequences $(b_t)_{t\in\mathbb{Z}}$ and $(c_t)_{t\in\mathbb{Z}}$ from (6.5). As in Example 6.2.7, the assumptions of Theorem 6.2.5 hold with $\lambda^* = 0$. The transversality condition is $g_{11} = \frac{4}{3} \neq 0$ and $D^2_{(1,2)} f_t(0,0,0) \equiv 0$ on $\mathbb{Z}$ implies $g_{20} = 0$, whereas

$$D^3_{(1,2)} f_t(0,0,0)\zeta^3 = \begin{pmatrix} 0 \\ 6\delta\zeta_1^3 \end{pmatrix} \quad \textit{for all } t \in \mathbb{Z} \textit{ and } \zeta \in \mathbb{R}^2$$

leads to $g_{30} = 4\delta \neq 0$. Having this available, one gets $\frac{g_{30}}{g_{11}} = 3\delta$. By Corollary 6.2.6(b), one deduces a pitchfork bifurcation of the trivial solution to (6.10) at $\lambda^* = 0$ being subcritical for $\delta > 0$ and supercritical for $\delta < 0$. We confirm this result using the general solution φ_λ to (6.10). As above, the first component is (6.6), which gives for the second component that

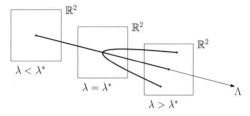

Figure 6.6 Supercritical pitchfork bifurcation from Example 6.2.8 with $\lambda^* = 0$ for different parameters λ: initial values (ξ, η) yielding solutions $\varphi_\lambda(\cdot, 0, \xi, \eta)$ of (6.10) being homoclinic to 0

$$\varphi_\lambda^2(t, 0, \xi, \eta) = \begin{cases} 2^t \left(\eta + \frac{8\delta}{15}\xi^3 + \frac{2\lambda}{3}\xi\right) + o(1) & as\ t \to \infty, \\ 2^{-t} \left(\eta - \frac{2\delta}{15}\xi^3 - \frac{4\lambda}{3}\xi\right) + o(1) & as\ t \to -\infty. \end{cases}$$

This asymptotic representation shows that $\varphi_\lambda(\cdot, 0, \xi, \eta)$ is bounded if and only if $\eta = 0$ or $\xi^2 = -\frac{2}{\delta}\lambda$, $\eta = \frac{4}{15}\frac{(5\delta+16\lambda)}{\delta^2}\lambda^2$. Hence, this is the pitchfork solution bifurcation illustrated in Fig. 6.6.

The solution bifurcation patterns discussed in Theorems 6.2.3 and 6.2.5 are based on Hypothesis 6.2.2. We note that its assumption (6.3) is not robust under parameter variation. This causes the dichotomy spectrum to behave as illustrated in Fig. 5.2, i.e. to suddenly shrink for parameters $\lambda \neq \lambda^*$. Therefore, the resulting bifurcations are rather exceptional (see also Remark 5.1.6).

6.2.3 Shovel Bifurcation

A shovel bifurcation pattern as studied in Sect. 3.2.2 also occurs for nonautonomous difference equations (Δ_λ) in $\mathbb{R}^d$ with obvious modifications (e.g. change the stability boundary 0 to 1). Technically, along a given branch $\phi(\lambda)$ of bounded solutions, one needs to assume invertible Jacobians $D_1 f_t(\phi(\lambda)_t, \lambda)$ for $t \in \mathbb{Z}$, $\lambda \in \Lambda$ (for details, see [189]). While the 'shovel' exists, the solutions $\phi(\lambda)$ are nonhyperbolic in the sense that $1 \in \Sigma(\lambda) \setminus \Sigma_s(\lambda)$ holds, i.e. 1 is not contained in the surjectivity spectrum of the variational equation (Y_λ).

A simple illustration provides the linear difference equation in Example 1.2.4. Here, schematically the bifurcating family of bounded solutions corresponding to Fig. 3.5 fills the whole half plane left (subcritical case) and, respectively, right (supercritical case) of the critical parameter λ^*.

6.3 Bifurcation of Minimal Sets and Invariant Graphs

In the context of skew product flows, it is natural to concentrate on bifurcations of invariant graphs, although bifurcations of minimal sets are also observed here, albeit as a less general phenomenon. For connections between minimal sets, minimal strips and pinched graphs (for minimally forced monotone interval maps), we refer the reader to [221].

Let $\theta : \mathbb{Z} \times \Omega \to \Omega$ be a dynamical system on a compact metric space Ω. We equip Ω with the Borel σ-algebra and an invariant measure μ. This section focuses on scalar nonautonomous difference equations

$$x_{t+1} = f(\theta(t, \omega), x_t, \lambda) \qquad (\Delta_\lambda^\omega)$$

with a right hand side $f : \Omega \times I \times \Lambda \to \mathbb{R}$, where $I, \Lambda \subseteq \mathbb{R}$ are intervals.

6.3.1 Fold Bifurcations of Invariant Graphs

In this subsection, the effect of external forcing on the fold bifurcation pattern of interval maps will be studied. The results can be viewed as discrete counterparts to results discussed in Sect. 3.3. The problem will be formulated for skew product flows over a compact metric space Ω, and the bifurcating objects will be invariant graphs. We also study bifurcations of minimal sets.

We first aim at generalising Example 1.2.6 by a criterion for the occurrence of fold bifurcations of invariant graphs in the context of nonautonomous monotone C^2 interval maps. The skew product flow generated by (Δ_λ^ω) is denoted as $(\theta, \varphi_\lambda)$.

Hypothesis 6.3.1 *Let $I \subset \mathbb{R}$ be a compact interval and $\lambda \in [0, 1]$. We suppose that there exist continuous functions $\gamma^-, \gamma^+ : \Omega \to I$ with $\gamma^- < \gamma^+$ such that for all $\lambda \in [0, 1]$ and $\omega \in \Omega$, the following hold:*

(i) *There exist two distinct continuous (θ, φ_0)-invariant graphs and no (θ, φ_1)-invariant graph in the region $\Gamma := \{(\omega, \xi) \in \Omega \times I : \gamma^-(\omega) \leq \xi \leq \gamma^+(\omega)\}$.*

(ii) $f(\omega, \gamma^\pm(\omega), \lambda) \geq \gamma^\pm(\theta(1, \omega))$.

(iii) *The partial derivatives $D_2^i f, D_3 f : \Omega \times I \times [0, 1] \to \mathbb{R}$ exist as continuous functions for $i \in \{0, 1, 2\}$ and*

$$D_2 f(\omega, \xi, \lambda) > 0, \qquad D_2^2 f(\omega, \xi, \lambda) > 0 \qquad and \qquad D_3 f(\omega, \xi, \lambda) > 0$$

for all $\xi \in \Gamma(\omega)$.

The following result is taken from [5, Theorem 6.1].

Theorem 6.3.2 (Subcritical Fold Bifurcation of Invariant Graphs) *Suppose that Hypothesis 6.3.1 holds. Then there exists a unique critical parameter $\lambda^* \in (0, 1)$ such that the following hold:*

(a) *For $\lambda < \lambda^*$, there exist two continuous $(\theta, \varphi_\lambda)$-invariant graphs $\phi_\lambda^- < \phi_\lambda^+$ in Γ. For any θ-invariant measure μ, they are hyperbolic in terms of the Lyapunov exponents $\chi_\mu(\phi_\lambda^-) < 1 < \chi_\mu(\phi_\lambda^+)$.*

(b) *For $\lambda = \lambda^*$, either there exists exactly one continuous $(\theta, \varphi_\lambda)$-invariant graph ϕ^* in Γ, or there exist two $(\theta, \varphi_\lambda)$-invariant graphs $\phi_\lambda^- \leq \phi_\lambda^+$ in Γ, with ϕ_λ^- lower and ϕ_λ^+ upper semi-continuous and*

$$\inf_{\omega \in \Omega} \left(\phi_\lambda^+(\omega) - \phi_\lambda^-(\omega)\right) = 0.$$

If μ is a θ-invariant measure, then in the first case $\chi_\mu(\phi^*) = 1$. In the second case $\phi_\lambda^-(\omega) = \phi_\lambda^+(\omega)$ μ-a.s. implies $\chi_\mu(\phi_\lambda^\pm) = 1$, whereas $\phi_\lambda^-(\omega) < \phi_\lambda^+(\omega)$ μ-a.s. implies hyperbolicity $\chi_\mu(\phi_\lambda^-) < 1 < \chi_\mu(\phi_\lambda^+)$.

(c) For $\lambda > \lambda^*$, there do not exist $(\theta, \varphi_\lambda)$-invariant graphs in Γ.

We note that via coordinate changes, analogous versions of the above result can be derived in case of concave fibre maps or decreasing behaviour in the bifurcation parameter λ, i.e.

$$D_2^2 f(\omega, \xi, \lambda) < 0 \quad \text{or} \quad D_3 f(\omega, \xi, \lambda) < 0 \quad \text{for all } \xi \in \Gamma(\omega).$$

In the special case of Example 1.2.6, the base flow θ is given by an irrational rotation and thus minimal, and the bifurcating objects are invariant graphs, or equivalently, the minimal sets defined by the corresponding graphs $\phi_\lambda^\pm$.

Remark 6.3.3 (Restricted Base Space) *If $\Omega_0 \subseteq \Omega$ denotes a compact and θ-invariant subset, then Theorem 6.3.2 also applies to the restricted skew product flow $(\theta, \varphi_\lambda)|_{\mathbb{Z} \times \Omega_0 \times I}$, and one obtains a new critical parameter*

$$\lambda_{\Omega_0}^\star = \sup \left\{ \lambda \in [0, 1] \, \middle| \, \begin{array}{l} \text{for all } \lambda' < \lambda \text{ there exist two uniformly} \\ \text{separated invariant graphs for } (\theta, \varphi_{\lambda'})|_{\mathbb{Z} \times \Omega_0 \times I} \end{array} \right\}.$$

In this case, invariant graphs defined over such subsets of Ω may still exist after the critical parameter λ^* and may continue to bifurcate as described by Theorem 6.3.2.

We now demonstrate that by choosing a base flow θ with more complicated dynamics, it is possible to construct a more complex bifurcation pattern, where an uncountable number of bifurcations take place, over a bifurcation interval $[\lambda^*, \hat\lambda^*]$.

Example 6.3.4 (Quasi-Periodic Fold Bifurcation) *On $\Omega = \mathbb{R}^2/\mathbb{Z}^2$, we consider the two-dimensional base flow $\theta : \mathbb{Z} \times \Omega \to \Omega$ generated by the mapping*

$$\theta_0(\omega_1, \omega_2) := \begin{pmatrix} \omega_1 + \frac{1}{2}\sin(2\pi(\omega_2 + \frac{1}{2}\sin 2\pi\omega_1)) \\ \omega_2 + \frac{1}{2}\sin(2\pi\omega_1) \end{pmatrix}.$$

It has both an uncountable number of ergodic invariant measures and of minimal sets, since its rotation set has nonempty interior (see [114, Section 5] for a discussion). Moreover, it exhibits four elliptic islands, centred around the points of the two two-periodic orbits

$$\Omega_1 = \left\{ (\tfrac{1}{4}, \tfrac{1}{4}), (\tfrac{3}{4}, \tfrac{3}{4}) \right\} \quad \text{and} \quad \Omega_2 = \left\{ (\tfrac{1}{4}, \tfrac{3}{4}), (\tfrac{3}{4}, \tfrac{1}{4}) \right\}.$$

For parameters $\lambda \in [0, 1]$, consider the nonautonomous difference equation

$$x_{t+1} = \arctan(\alpha x_t) - \tfrac{1}{2}\left(\sin(2\pi\theta_1(t, \omega_1, \omega_2))\sin(2\pi\theta_2(t, \omega_1, \omega_2) + 1))\right) - 2\lambda$$

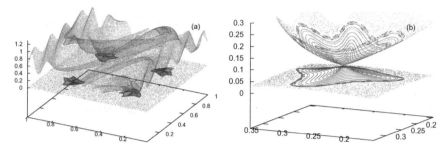

Figure 6.7 (a) Repelling (red) and attracting (green) invariant graphs at the critical parameter $\lambda^* \approx 0.1855650809$. (b) Closer view of the two invariant graphs over the islands centred at the two-periodic point $(\frac{1}{4}, \frac{1}{4})$

yielding a skew product flow $(\theta, \varphi_\lambda)$. We set $\alpha = 100$, as in the non-smooth case of Example 1.2.6, where a strange non-chaotic attractor–repeller pair emerges. It is easily shown that $(\theta, \varphi_\lambda)$ satisfies the assumptions of Theorem 6.3.2 and therefore undergoes a fold bifurcation. Note that the θ-dependent term

$$-\tfrac{1}{2} \sin(2\pi\theta_1(t, \omega_1, \omega_2)) \sin(2\pi\theta_2(t, \omega_1, \omega_2))$$

attains its global minimum exactly on the minimal set Ω_1 (respectively, its global maximum on the minimal set Ω_2), which implies that Ω_1 is precisely the set of points on which the two invariant graphs touch at the bifurcation point. Here invariant graphs defined on subsets of Ω continue to exist after the bifurcation has taken place, and they continue to bifurcate over an interval $[\lambda^*, \hat{\lambda}^*] = [\lambda^\star_{\Omega_1}, \lambda^\star_{\Omega_2}]$. In other words, Ω_1 is the minimal set on which the first bifurcation occurs, and Ω_2 is the minimal set on which the last bifurcation occurs, that is, $\lambda^\star_{\Omega_1} = \lambda^* < \lambda^\star_{\Omega_0} < \lambda^\star_{\Omega_2}$ for all minimal sets $\Omega_0 \neq \Omega_1, \Omega_2$. Figure 6.7 shows the two invariant graphs at the bifurcation point, together with a magnified view of the islands centred at the two-periodic point $(\frac{1}{4}, \frac{1}{4})$.

Similarly to Example 1.2.6, there exists a third invariant graph below $\Omega \times \{0\}$ that is continuous and attracting and that persists throughout the whole parameter range. Once the bifurcation has taken place over a minimal set $M \subset \Omega$, this graph attracts all trajectories in $M \times [-5, 2]$. Consequently, after the first bifurcation has taken place (over the minimal set Ω_1), all the invariant circles in the elliptic island, starting in the middle and moving outwards, also bifurcate. When the outer boundary of the two elliptic islands containing Ω_1 is reached, the complement of the elliptic islands (the chaotic region in the sense of [114]) drops instantaneously. Finally, the invariant circles over the remaining two elliptic islands drop down one by one, in reverse order, moving inwards from the outside and the last bifurcation takes place over the minimal set Ω_2.

In Fig. 6.8, the bifurcation over one of the invariant circles of the elliptic island is shown. Although embedded in dimension two, the underlying dynamics are just those of an irrational rotation. Consequently, from a qualitative point of view, the

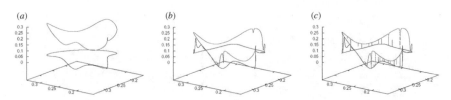

Figure 6.8 Closer view of two invariant circles above the island centred at the point $\left(\frac{1}{4}, \frac{1}{4}\right)$

behaviour of the system is the same as in Example 1.2.6: the non-uniform approach of the invariant circles can be observed, which is typical for the creation of strange non-chaotic attractors and repellers at the bifurcation point.

6.3.2 Fold Bifurcation in the Logistic Equation

The above Theorem 6.3.2 is formulated locally in the sense that attention is restricted to a region where the bifurcation of invariant graphs is taking place. One advantage of this formulation is that it allows to describe local bifurcations appearing in forced noninvertible interval maps. This idea is used to describe the creation of three-periodic invariant graphs in the quasi-periodically forced logistic map.

As a motivation, we consider the well-known logistic family

$$x_{t+1} = \lambda x_t (1 - x_t) \tag{6.11}$$

studied for instance in [11, Section 4.5] and [164]. Its three-periodic orbit is born via a supercritical fold bifurcation at the critical parameter $\lambda^* = 1 + \sqrt{8}$. In more detail, for this parameter it has a unique three-periodic orbit $\{\xi_1, \xi_2, \xi_3\}$ with $\xi_1 < \xi_2 < \xi_3$. There exist $\delta > 0$ and intervals $I_1, I_2, I_3 \subset [0, 1]$ containing ξ_1, ξ_2, ξ_3, respectively, so that the following hold:

(a) If $\lambda \in (\lambda^* - \delta, \lambda^*)$, then (6.11) has no three-periodic points in $I_1 \cup I_2 \cup I_3$.

(b) If $\lambda = \lambda^*$, then each of the intervals I_1, I_2, I_3 contains a unique three-periodic point, ξ_1, ξ_2, ξ_3 respectively.

(c) If $\lambda \in (\lambda^*, \lambda^* + \delta)$, then (6.11) has exactly two three-periodic points in I_i, denoted by $\xi_i^-(\lambda) < \xi_i^+(\lambda)$.

We now turn to a quasi-periodically forced version of (6.11). Let $\varepsilon \geq 0$, and consider a bifurcation parameter $\lambda \in [0, \frac{4}{1+\varepsilon}]$. For some $p \in \mathbb{N}$, consider the non-autonomous difference equation

$$x_{t+1} = \lambda(1 + \varepsilon \cos(2\pi\theta(t, \omega))^p)x_t(1 - x_t), \tag{6.12}$$

driven by an irrational rotation of the circle $\Omega = \mathbb{R}/\mathbb{Z}$ given by

$$\theta : \mathbb{Z} \times \Omega \to \Omega, \qquad \theta(t, \omega) := \omega + t\alpha \pmod{1},$$

where α is an irrational number. This difference equation generates a family of skew product flows $(\theta, \varphi_{\varepsilon,\lambda})$, which was studied in [167, 170, 197]. In particular, (6.12) has been described as a prototypical example for the birth of strange non-chaotic attractors (see the Remarks to Sect. 6.3).

The following result due to T.Y. Nguyen, T.S. Doan, T. Jäger and S. Siegmund [170, Theorem 4] is an application of Theorem 6.3.2 and requires yet another terminology: a graph $\phi : \Omega \to I$ is called T-*periodic* if T is the smallest positive integer satisfying

$$\varphi_{\varepsilon,\lambda}(T, \omega, \phi(\omega)) = \phi(\theta(T, \omega)) \quad \text{for all } \omega \in \Omega$$

and an orbit of T-periodic graphs is a collection $\{\varphi_{\varepsilon,\lambda}(s, \omega, \phi(\omega)) : 1 \le s \le T\}$. We can now describe a local fold bifurcation of three-periodic graphs in the quasi-periodically forced difference equation (6.12).

Theorem 6.3.5 (Supercritical Fold Bifurcation in (6.12)) *Consider the family of skew product flows* $(\theta, \varphi_{\varepsilon,\lambda})$ *generated by* (6.12). *Then there exist* $\delta^* > 0$, $\varepsilon^* > 0$ *and* $\eta > 0$ *and a function* $\lambda^* : [0, \varepsilon^*] \to [0, \frac{4}{1+\varepsilon^*}]$, *such that the following statements hold for each* $\varepsilon \in [0, \varepsilon^*]$:

(a) *If* $\lambda \in (\lambda^*(\varepsilon) - \delta^*, \lambda^*(\varepsilon))$, *then there does not exist a three-periodic graph in the sets* $K_1(\eta)$, $K_2(\eta)$, $K_3(\eta)$, *where*

$$K_i(\eta) := (\mathbb{R}/\mathbb{Z}) \times [\xi_i^-(\lambda^* + \eta), \xi_i^+(\lambda^* + \eta)] \quad \text{for all } i \in \{1, 2, 3\} .$$

(b) *If* $\lambda = \lambda^*(\varepsilon)$, *then each of the sets* $K_1(\eta)$, $K_2(\eta)$, $K_3(\eta)$ *contains exactly one continuous three-periodic graph (together forming a three-periodic orbit) or a pair of pinched semi-continuous three-periodic graphs (one upper semi-continuous and one lower semi-continuous in each* $K_i(\eta)$, *forming two orbits of three-periodic graphs, respectively).*

(c) *If* $\lambda \in (\lambda^*(\varepsilon), \lambda^*(\varepsilon) + \delta^*)$, *then there are two continuous three-periodic graphs in* $K_i(\eta)$ *(forming two orbits of three-periodic graphs), denoted by* $\phi_{\lambda,\varepsilon,i}^-$ *and* $\phi_{\lambda,\varepsilon,i}^+$ *with* $\phi_{\lambda,\varepsilon,i}^-(\omega) < \phi_{\lambda,\varepsilon,i}^+(\omega)$ *for all* $\omega \in \mathbb{R}/\mathbb{Z}$ *and* $i \in \{1, 2, 3\}$.

We close with an illustration of Theorem 6.3.5.

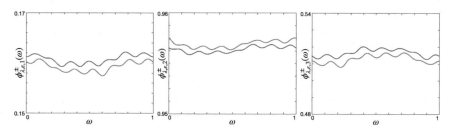

Figure 6.9 Three-periodic repelling graphs of $\phi_{\lambda,\varepsilon,i}^+$ (red) and attracting graphs of $\phi_{\lambda,\varepsilon,i}^-$ (green) for the parameter $\lambda = 3.828580 > \lambda^*$

Example 6.3.6 *The values $p = 9$, $\varepsilon = 0.00006$ and $\alpha = \frac{\sqrt{5}-1}{2}$ in (6.12) yield the critical parameter $\lambda^* \approx 3.828529$. Then Fig. 6.9 shows the attracting and repelling orbits of three-periodic graphs that exist after the bifurcation has taken place.*

Remarks

An early contribution to nonautonomous bifurcations in difference equations is [198], relying on attractor bifurcation. Also in discrete time, classical autonomous transcritical and pitchfork bifurcations can be understood as attractor bifurcations (for this, see [198, Example 5.3 and, respectively, Remark 6.2(vi)].

It seems interesting to obtain a nonautonomous, discrete-time version of the attractor bifurcation studied in [163, pp. 114ff, Section 5.2]. Also [135] might be helpful to relate attractor and solution bifurcations, since (global) attractors consist of bounded entire solutions [186, p. 17, Corollary 1.3.4].

Although the proof techniques from [150–152] do not immediately carry over to difference equations, the resulting solution bifurcations can be illustrated in discrete time. For this purpose, T. Hüls [106] provides a detailed bifurcation analysis in terms of the explicitly solvable model

$$x_{t+1} = \frac{\lambda x_t}{(1 + \frac{a_t q}{\lambda} x_t^q)^{1/q}},$$

which exhibits both nonautonomous transcritical ($q = 1$ yields the *Beverton–Holt equation*) and pitchfork bifurcations ($q = 2$). They are also related to the associated attractor bifurcation concepts in [198]. A similar planar model was studied in [110].

The alternative approach via solution bifurcations arose later from [187, 189]. In this context we point out that the results from Sects. 6.2.1 and 6.2.2 immediately generalise to difference equations in Banach spaces. Under the assumption $\lim_{t \to \pm\infty} f_t(0, \lambda) = 0$ for all $\lambda \in \Lambda$, one can show that the bifurcating solutions in Theorems 6.2.3 and 6.2.5 are indeed homoclinic to 0, i.e. contained in the space of sequences converging to 0 in both directions. Related results on solutions in form of limit-zero sequences are presented in [187, 190]. Extensions using topological arguments can be found in [182, 219]. Persistence results addressing changes in the above solution bifurcations and their bifurcation diagrams under perturbation are investigated in [191]. The essential linear theory for shovel bifurcations is based on exponential trichotomies [181].

Almost Periodic Bifurcation Bifurcation of almost periodic solutions is studied in [98], which is a discrete-time version of results from [142]. A more recent contribution is [208] studying the Ricker equation from ecology under almost periodic forcing. We refer to [89] for a result on almost periodic variational equations.

Further results on fold bifurcation in quasi-periodically forced interval maps are due to [86, 88]. A pitchfork bifurcation under quasi-periodic forcing can be found in [125].

Strange Non-Chaotic Attractors Strange non-chaotic attractors (SNAs) are unusual objects which combine complicated geometry with non-chaotic dynamics. Independently discovered by Herman [100] and Grebogi et al. [93] in quasi-periodically forced systems, SNAs have been studied extensively in various systems [90, 130, 165, 176, 197]. Non-continuous invariant graphs with negative Lyapunov exponents are usually referred to as examples of strange non-chaotic attractors. In nonautonomous quasi-periodically forced systems, these are often created during the collision of invariant curves, as seen in Example 1.2.6, where a strange non-chaotic attractor–repeller pair emerges at the bifurcation point. Attractor and repeller are interwoven in such a way that they have the same topological closure. This particular route for the creation of SNAs has been observed quite frequently, for example in [5, 90, 168].

Chapter 7
Reduction Techniques

Also in discrete time we pursue two possibilities to simplify a given bifurcation problem. Our presentation is parallel to Chap. 4.

On the one hand, we give a rather general description of nonautonomous locally invariant manifolds. Due to their variation in time, we speak of *fibre bundles*, which are graphs of functions defined over the spectral manifolds. Among them, centre bundles are a special case, as are the stable and unstable bundles previously mentioned in Sect. 5.3.2. In order to make the reduction principle applicable, we provide a detailed approach to obtain Taylor approximations of invariant fibre bundles and indicate their usefulness in bifurcation theory.

On the other hand, normal form theory is an essential tool for the bifurcation analysis of autonomous systems [148, 228]. Although we do not have applications for nonautonomous equations at hand, we nevertheless review the corresponding theory. Yet, already scalar equations demonstrate that no simplifications can be expected in the nonhyperbolic situation.

We focus on problems defined on discrete intervals $\mathbb{I}$ being unbounded below. Consider a parameter-free nonautonomous difference equation

$$x_{t+1} = f_t(x_t) \tag{Δ}$$

with a C^m-right hand side $f : \mathcal{D} \to \mathbb{R}^d$, $m \geq 2$, on an open domain $\mathcal{D} \subseteq \mathbb{I} \times \mathbb{R}^d$. We suppose that (Δ) admits a fixed reference solution $\phi^* = (\phi_t^*)_{t \in \mathbb{I}}$ satisfying the inclusion $\mathcal{B}_R(\phi^*) \subseteq \mathcal{D}$ for some $R > 0$. Let us denote the corresponding variational difference equation by

$$x_{t+1} = Df_t(\phi_t^*)x_t$$

and $\Sigma_{\mathbb{I}}(\phi^*) = \bigcup_{i=1}^{n}[a_i, b_i]$ is the corresponding dichotomy spectrum.

© The Author(s), under exclusive license to Springer Nature Switzerland AG 2023
V. Anagnostopoulou et al., *Nonautonomous Bifurcation Theory*, Frontiers
in Applied Dynamical Systems: Reviews and Tutorials 10,
https://doi.org/10.1007/978-3-031-29842-4_7

7.1 Centre Fibre Bundles

This section introduces a nonautonomous counterpart to centre manifolds, so-called *centre fibre bundles*. It serves as a dynamically meaningful tool to reduce the dimension of bifurcation problems.

We pass over to the equation of perturbed motion

$$x_{t+1} = Df_t(\phi_t^*)x_t + F_t(x_t) \tag{7.1}$$

and assume that the C^m-nonlinearity $F_t(x) := f_t(x + \phi_t^*) - f_t(\phi_t^*) - Df_t(\phi_t^*)x$ fulfils the limit relation

$$\lim_{x \to 0} DF_t(x) = 0 \quad \text{uniformly in } t \in \mathbb{I}. \tag{7.2}$$

We stick to a rather general framework and choose reals $0 < \beta_+ < \beta_-$ in a gap of the dichotomy spectrum $\Sigma_\mathbb{I}(\phi^*)$ of the variational equation (Y), i.e.

$$\Sigma_\mathbb{I}(\phi^*) \cap (\beta_+, \beta_-) = \emptyset.$$

The invariant projector associated with the exponential dichotomy of the scaled equation

$$x_{t+1} = \gamma^{-1} Df_t(\phi_t^*)x_t$$

with $\gamma \in (\beta_+, \beta_-)$ is denoted as P_t and $\mathcal{V}^-$ is the corresponding unstable set.

We aim at describing a nonautonomous counterpart of an invariant manifold for (Δ) along ϕ^*, and we do so by equivalently finding an invariant manifold along the trivial solution of (7.1). Let $U \subseteq \mathbb{R}^d$ be an open convex neighbourhood of 0. Suppose the functions $c_t : U \to \mathbb{R}^d$ for $t \in \mathbb{I}$ are continuously differentiable with

$$c_t(0) \equiv 0 \quad \text{on } \mathbb{I}, \quad c_t(x) = c_t((\mathrm{id} - P_t)x) \in R(P_t) \quad \text{for all } t \in \mathbb{I} \text{ and } x \in U,$$

$$\lim_{x \to 0} Dc_t(x) = 0 \quad \text{uniformly in } t \in \mathbb{I}. \tag{7.3}$$

Under these conditions, the graph

$$\phi^* + \mathcal{C} := \left\{ (\tau, \phi_\tau^* + \xi + c_\tau(\xi)) \in \mathbb{I} \times \mathbb{R}^d : \xi \in N(P_\tau) \cap U \right\}$$

is called *locally invariant fibre bundle* for the solution ϕ^* to (Δ) (see Fig. 7.1), provided that

$$(t_0, x_0) \in \phi^* + \mathcal{C} \quad \Rightarrow \quad (t, \varphi(t, t_0, x_0)) \in \phi^* + \mathcal{C} \quad \text{for all } t_0 \leq t$$

holds, as long as the general solution φ to (Δ) satisfies $\varphi(t, t_0, x_0) \in \phi^* + U$.

Specifically, one speaks of a C^m-*fibre bundle*, if the derivatives $D^n c_t$ exist and are continuous for $1 \leq n \leq m$. The existence of locally invariant fibre bundles follows, for instance, from [194, Theorem 3.2].

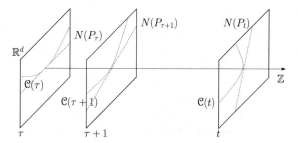

Figure 7.1 Fibres $\mathcal{C}(t)$ of an invariant bundle $\mathcal{C} \subseteq \mathbb{I} \times \mathbb{R}^d$ along the trivial solution being smooth curves tangential to the kernels $N(P_t) = \mathcal{V}^-(t)$ for all $t \in \mathbb{I} \subseteq \mathbb{Z}$

Theorem 7.1.1 (Existence of Locally Invariant Bundles) *There exist reals $\rho_0 > 0$ and $\gamma_0, \ldots, \gamma_m \geq 0$ such that the following holds with $U = B_{\rho_0}(0)$. If the spectral gap condition*

$$\beta_+ < \beta_-^m$$

is satisfied, then the solution ϕ^* to (Δ) possesses a locally invariant C^m-fibre bundle $\mathcal{C}$ with the following properties:

(a) *The corresponding mappings $c_t : U \to \mathbb{R}^d$, $t \in \mathbb{I}$, satisfy*

$$\|D^n c_t(x)\| \leq \gamma_n \quad \text{for all } t \in \mathbb{I}, x \in U \text{ and } n \in \{0, \ldots, m\} . \qquad (7.4)$$

(b) *If the right hand side f_t and the solution ϕ^* are T-periodic, then*

$$c_{t+T} = c_t, \qquad \mathcal{C}(t) = \mathcal{C}(t+T) \quad \text{for all } t \in \mathbb{Z}.$$

In particular for autonomous difference equations (Δ) and constant solutions ϕ^* there is a C^m-mapping $c : U \to \mathbb{R}^d$ with $c \equiv c_t$ on $\mathbb{I}$, i.e. $\mathcal{C}$ has constant fibres $\{\phi^* + \xi + c(\xi) \in \mathbb{R}^d : \xi \in N(P) \cap U\}$, where $P \in \mathbb{R}^{d \times d}$ is the spectral projector corresponding to the eigenvalues of $Df(\phi^*)$ in $B_{\beta_+}(0)$.

Until now, besides being disjoint from the dichotomy spectrum $\Sigma_{\mathbb{I}}(\phi^*)$, we made no further assumption on the growth rates (β_+, β_-) yet. In the situation where 1 is contained in the dominant spectral interval $[a_n, b_n]$ and β_- and β_+ are chosen from the gap just left of it, then $\mathcal{C}$ is denoted as *centre fibre bundle* of ϕ^*.

The usefulness of centre fibre bundles is due to the fact that they allow a dimension reduction in such critical stability situations. The following reduction principle is taken from [186, p. 267, Theorem 4.6.14].

Theorem 7.1.2 (Reduction Principle) *Suppose that $\mathbb{I} = \mathbb{Z}$ and $1 \in [a_n, b_n]$ hold. A solution ϕ^* of (Δ) is stable (uniformly stable, asymptotically stable, uniformly asymptotically stable, exponentially stable, uniformly exponentially stable or unstable), if and only if the trivial solution to the reduced equation*

$$x_{t+1} = Df_t(\phi_t^*)x_t + (\text{id} - P_{t+1})F_t(x_t + c_t(x_t)) \qquad (7.5)$$

in the set $\mathcal{V}^-$ has the respective stability property.

Being a difference equation in $\mathcal{V}^-$, the dimension of the reduced equation (7.5) is given by the multiplicity of the dominant spectral interval.

In order to make use of (7.5), at least Taylor coefficients of the nonlinearity are required. In order to describe the procedure to obtain them, we return to the general situation where (β_+, β_-) is chosen from an arbitrary gap in $\Sigma_{\mathbb{I}}(\phi^*)$. Indeed, we are interested in local approximations of the mappings $c_t : U \to \mathbb{R}^d$, $t \in \mathbb{I}$, describing a C^m-invariant fibre bundle for the solution ϕ^* to (Δ). Taylor's theorem [229, p. 148, Theorem 4.A] together with (7.3) implies the representation

$$c_t(x) = \sum_{n=2}^{m} \frac{1}{n!} c_t^n x^{(n)} + R_t^m(x) \qquad (7.6)$$

with coefficients $c_t^n \in L_n(\mathbb{R}^d)$ given by $c_t^n := D^n c_t(0)$ and a remainder R_t^m satisfying $\lim_{x \to 0} \frac{1}{\|x\|^m} R_t^m(x) = 0$ uniformly in $t \in \mathbb{I}$. Due to (7.4), the sequences $(c_t^n)_{t \in \mathbb{I}}$ are bounded with $\|c_t^n\| \le \gamma_n$ for $t \in \mathbb{I}$ and $2 \le n \le m$ with $\gamma_2, \ldots, \gamma_m \ge 0$. We need further preparations and it is convenient to write $Q_t := \mathrm{id} - P_t$.

(i) We introduce $C_t : U \to \mathbb{R}^d$, $C_t(x) := Q_t x + c_t(x)$, satisfying

$$DC_t(0) \overset{(7.3)}{=} Q_t \quad \text{and} \quad D^n C_t(0) = D^n c_t(0) \quad \text{for all } t \in \mathbb{I}$$

and $n \in \{2, \ldots, m\}$. Hence, for the derivatives $C_t^n := D^n C_t(0)$, we have

$$\|C_t^1\| \le K \quad \text{and} \quad \|C_t^n\| \overset{(7.4)}{\le} \gamma_n \quad \text{for all } n \in \{2, \ldots, m\}.$$

(ii) We define $g_t(x) := Q_{t+1} [Df_t(\phi_t^*)x + F_t(Q_t x + c_t(x))]$ and note that the partial derivatives $g_t^n := D^n g_t(0)$ are given by

$$g_t^1 x_1 \overset{(5.1)}{=} Df_t(\phi_t^*) Q_t x_1,$$

$$g_t^n x_1 \cdots x_n = \sum_{l=2}^{n} \sum_{(N_1,\ldots,N_l) \in P_l^<(n)} Q_{t+1} D^l F_t(0) C_t^{\#N_1}|_{Q_t} x_{N_1} \cdots C_t^{\#N_l}|_{Q_t} x_{N_l}$$

for all $x_1, \ldots, x_n \in \mathbb{R}^d$ and $n \in \{2, \ldots, m\}$, see (7.2)–(7.3). The ordered partitions $P_l^<(n)$ were introduced in Sect. 2.3.

Given multi-linear mappings $X \in L_n(\mathbb{R}^d)$ and $T \in L(\mathbb{R}^d)$, we introduce the notation $X|_T x_1 \cdots x_n := X(Tx_1, \ldots, Tx_n)$ for $x_1, \ldots, x_n \in \mathbb{R}^d$. In [194], we establish that each Taylor coefficient sequence c^n is a solution to the *homological equation* for $\mathcal{C}$ given by

$$X_{t+1}|_{Df_t(\phi_t^*)Q_t} = Df_t(\phi_t^*) X_t|_{Q_t} + H_t^n|_{Q_t},$$

which is a linear difference equation in the space $L_n(\mathbb{R}^d)$ with inhomogeneities $H_t^n \in L_n(\mathbb{R}^d)$, given by

$$H_t^n x_1 \cdots x_n := P_{t+1}\Big[D^n F_t(0)|_{Q_t} x_1 \cdots x_n$$

$$+ \sum_{l=2}^{n-1} \sum_{(N_1,\dots,N_l)\in P_t^{<}(n)} \big(D^l F_t(0) C_t^{\# N_1}|_{Q_t} x_{N_1} \cdots C_t^{\# N_l}|_{Q_t} x_{N_l}$$

$$- c_{t+1}^l g_t^{\# N_1}|_{Q_t} x_{N_1} \cdots g_t^{\# N_l}|_{Q_t} x_{N_l}\big)\Big].$$

Obviously, one has $H_2^k = P_{t+1} D^2 F_t(x)|_{Q_t}$, and for $n \in \{3,\dots,m\}$, the values H_t^n only depend on the sequences $c^2,\dots,c^{n-1}$. This leads to the following result taken from [194, Theorem 4.2(b)].

Theorem 7.1.3 (Taylor Expansion of Invariant Fibre Bundle) *Let $t \in \mathbb{I}$. The coefficients $c_t^n \in L_n(\mathbb{R}^d)$, $2 \leq n \leq m$, in the Taylor expansion (7.6) of the mapping $c_t : U \to \mathbb{R}^d$ can be determined recursively from the* Lyapunov–Perron *sums*

$$c_t^n = \sum_{s=-\infty}^{t-1} \Phi(t, s+1) H_s^n|_{\Phi(s,t)Q_t} \quad \text{for all } t \in \mathbb{I} \text{ and } 2 \leq n \leq m.$$

For example, illustrating the above theorem, we refer to a discrete epidemic model studied in [194, Examples 5.1 and 5.5].

In order to apply centre fibre bundles in bifurcation theory for (Δ_λ), one proceeds as in Sect. 4.1, but now with the *augmented difference equation*

$$\begin{cases} x_{t+1} = f_t(x_t, \lambda_t), \\ \lambda_{t+1} = \lambda_t. \end{cases}$$

7.2 Normal Forms

We sketch the nonautonomous normal form theory of S. Siegmund [217], which parallels the continuous time situation described in Sect. 4.2. For $\mathbb{I} = \mathbb{Z}$, we suppose a variational equation (Y) with bounded growth, as well as

(a) For all $t \in \mathbb{Z}$, the right hand sides $f_t|_{B_R(\phi_t^*)}$ are C^m-diffeomorphisms onto $f_t(B_R(\phi_t^*))$ for some $m > 1$

(b) $\sup_{t\in\mathbb{Z}} \|D^j f_t(\phi_t^*)\| < \infty$ for all $2 \leq j \leq m$.

Then the nonautonomous difference equation (Δ) is locally C^m-equivalent between the solution ϕ^* and the trivial solution of a difference equation

$$x_{t+1} = B_t x_t + g_t(x_t) \tag{7.7}$$

with the following properties:

(i) The linear part is block diagonal consisting of n blocks $B_t^j \in GL(\mathbb{R}^{d_j})$ with interval dichotomy spectra $\Sigma(B^j) = [a_j, b_j]$ for $1 \leq j \leq n$, and we have $d_1 + \dots + d_n = d$ and $\Sigma(\phi^*) = \Sigma(B)$.

(ii) The right hand side of (7.7) is a C^m-diffeomorphism on some $B_{r_0}(0) \subset \mathbb{R}^d$ for every $t \in \mathbb{Z}$ and $Dg_t(0) = 0$ for all $t \in \mathbb{Z}$.

Before stating a discrete time counterpart to Theorem 4.2.1, let us introduce the following arithmetics for intervals: for reals $0 < a \leq b$ and $0 < c \leq d$, we define the product $[a, b] \cdot [c, d] := [ac, bd]$ of intervals and their powers $[a, b]^j := [a^j, b^j]$ for $j \in \mathbb{N}$. The discrete time normal form theorem taken from [217, Theorem 11] shows that higher order coefficients in g_t can be eliminated.

Theorem 7.2.1 (Normal Form Theorem) *If* $1 \leq j \leq n$ *and a multi-index* $\alpha \in \mathbb{N}_0^n$ *with* $2 \leq |\alpha| \leq m$ *satisfies the* nonresonance *condition*

$$[a_j, b_j] \cap \prod_{i=1}^{n}[a_i, b_i]^{\alpha_i} = \emptyset,$$

then $D^\alpha \Pi_j g_t(0) \equiv 0$ *on* $\mathbb{Z}$ *with the projector* $\Pi_j \in \mathbb{R}^{d \times d}$ *onto* $\{0\} \times \mathbb{R}^{d_j} \times \{0\}$.

The next example explains the normal form result in the one-dimensional case.

Example 7.2.2 (Normal Form for One-Dimensional Equations) *Whenever* $d = 1$, *the scalar difference equation (7.1) of perturbed motion along* ϕ^* *has the Taylor expansion*

$$x_{t+1} = Df_t(\phi_t^*)x_t + \sum_{j=2}^{m} \frac{D^j f_t(\phi_t^*)}{j!}x_t^j + R_m(x_t).$$

The spectrum $\Sigma(\phi^*) = [a, b]$ *is a compact interval, whose boundary points are the Bohl exponents of the sequence* $(Df_t(\phi_t^*))_{t \in \mathbb{Z}}$. *The nonresonance condition to eliminate the* j-*th order term becomes* $[a, b] \cap [a^j, b^j] = \emptyset$. *In the nonhyperbolic case* $1 \in [a, b]$, *this is never satisfied and there is no hope for an algebraic simplification of* (Δ) *near* ϕ^*. *Otherwise, for* $j \in \{2, \ldots, m\}$, *there are two choices: in the uniformly exponentially stable case* $b < 1$, *the terms of order* $j > \frac{\ln a}{\ln b}$ *can be eliminated, and if* $1 < a$, *then terms of order* $j > \frac{\ln b}{\ln a}$ *can be eliminated.*

Remarks

Invariant Fibre Bundles A comprehensive approach to nonautonomous invariant manifolds, in discrete time we speak of *invariant fibre bundles*, is given in [186, pp. 187ff, Chapter 4]. In particular, by means of centre fibre bundles, one can reduce the dynamics of a critical difference equation to its essential part. For this reason, the reduced difference equation (7.5) has a critical linear part. Note that unless it is scalar, corresponding stability investigations require refined techniques. A simple criterion, which applies to scalar equations, is given in [194, Proposition 5.4].

Our Theorem 7.1.3 can be used to obtain local approximations of invariant fibre bundles based on Taylor coefficients. This is sufficient for a local stability and bifurcation analysis.

Normal Forms Our nonautonomous theory is due to [217]. In case of a discrete dichotomy spectrum $\Sigma(\phi^*) = \bigcup_{i=1}^{n} \{\lambda_j\}$, the nonresonance conditions reduce to $\{\lambda_j\} \cap \{\prod_{i=1}^{n} \lambda_i^{\alpha_i}\} = \emptyset$, which are precisely the classical nonresonance conditions $\lambda_j \neq \prod_{i=1}^{n} \lambda_i^{\alpha_i}$ from the autonomous theory [228, pp. 211, Section 2.2], where λ_j denote the eigenvalues of the linearisation in a fixed point.

Appendix

Our appendix consists of two parts, which both address uniform versus nonuniform stability. In the first one we precisely define the stability notions used throughout the text and describe exponential stability in terms of dynamical spectra. The second one aims to summarize properties of Bohl and Lyapunov exponents in order to provide some intuition for them.

A.1 Stability Theory

Let $\mathcal{D} \subseteq \mathbb{T} \times \mathbb{R}^d$ be nonempty (and open in case $\mathbb{T} = \mathbb{R}$). In this appendix we briefly introduce the stability notions needed in the preceding text and refer to the monographs [1, 24, 59, 72, 96, 101, 112, 153] for a more comprehensive treatment.

Consider the nonautonomous differential or difference equation

$$\dot{x} = f(t, x) \quad \text{or} \quad x_{t+1} = f(t, x_t), \tag{A.1}$$

with a right hand side $f : \mathcal{D} \to \mathbb{R}^d$ such that the existence and uniqueness of forward solutions is guaranteed; the general solution is denoted by φ. Let $\phi^* : \mathbb{I} \to \mathbb{R}^d$ be a solution to (A.1) on an interval $\mathbb{I} \subseteq \mathbb{T}$ unbounded above with $\mathcal{B}_R(\phi^*) \subseteq \mathcal{D}$ for some $R > 0$. This reference solution ϕ^* is called

- *Stable*, if for all $\varepsilon > 0$ $\tau \in \mathbb{I}$, there exists a $\delta > 0$ such that $\varphi(\cdot, \tau, \xi)$ exists and satisfies $\|\varphi(t, \tau, \xi) - \phi^*(t)\| < \varepsilon$ for all $\tau \leq t$, $\xi \in B_\delta(\phi^*(\tau))$

- *Asymptotically stable*, if it is stable and *attractive*, that is for all $\tau \in \mathbb{I}$ there is a $\rho > 0$ so that $\varphi(\cdot, \tau, \xi)$ exists and satisfies $\lim_{t \to \infty} \|\varphi(t, \tau, \xi) - \phi^*(t)\| = 0$ for all $\xi \in B_\rho(\phi^*(\tau))$

- *Exponentially stable*, if for all $\tau \in \mathbb{I}$ there exist $K \geq 1$ and in case

- $\mathbb{T} = \mathbb{R}$ real numbers $\alpha < 0$, $\delta > 0$ such that $\varphi(\cdot, \tau, \xi)$ exists and satisfies $\|\varphi(t, \tau, \xi) - \phi^*(t)\| \leq K e^{\alpha t} \|\xi - \phi^*(\tau)\|$ for all $\tau \leq t$, $\xi \in B_\delta(\phi^*(\tau))$

© The Author(s), under exclusive license to Springer Nature Switzerland AG 2023
V. Anagnostopoulou et al., *Nonautonomous Bifurcation Theory*, Frontiers in Applied Dynamical Systems: Reviews and Tutorials 10,
https://doi.org/10.1007/978-3-031-29842-4

○ $\mathbb{T} = \mathbb{Z}$ real numbers $\alpha \in (0,1)$, $\delta > 0$ such that $\varphi(\cdot, \tau, \xi)$ exists and satisfies $\|\varphi(t, \tau, \xi) - \phi^*(t)\| \leq K\alpha^t \|\xi - \phi^*(\tau)\|$ for all $\tau \leq t$, $\xi \in B_\delta(\phi^*(\tau))$

• *Uniformly stable* (on $\mathbb{I}$), if for all $\varepsilon > 0$ there exists a $\delta > 0$ such that $\varphi(\cdot, \tau, \xi)$ exists and satisfies $\|\varphi(t, \tau, \xi) - \phi^*(t)\| < \varepsilon$ for all $\tau \leq t$, $\xi \in B_\delta(\phi^*(\tau))$

• *Uniformly asymptotically stable* (on $\mathbb{I}$), if it is uniformly stable (on $\mathbb{I}$) and there is a $\rho > 0$ so that $\varphi(\cdot, \tau, \xi)$ exists and satisfies $\lim_{t \to \infty} \|\varphi(t, \tau, \xi) - \phi^*(t)\| = 0$ for all $\tau \in \mathbb{I}$, $\xi \in B_\rho(\phi^*(\tau))$

• *Uniformly exponentially stable* (on $\mathbb{I}$), if there exists $K \geq 1$ and in case

○ $\mathbb{T} = \mathbb{R}$ real numbers $\alpha < 0$, $\delta > 0$ such that $\varphi(\cdot, \tau, \xi)$ exists and satisfies $\|\varphi(t, \tau, \xi) - \phi^*(t)\| \leq Ke^{\alpha(t-\tau)} \|\xi - \phi^*(\tau)\|$ for all $\tau \leq t$, $\xi \in B_\delta(\phi^*(\tau))$

○ $\mathbb{T} = \mathbb{Z}$ real numbers $\alpha \in (0,1)$, $\delta > 0$ such that $\varphi(\cdot, \tau, \xi)$ exists and satisfies $\|\varphi(t, \tau, \xi) - \phi^*(t)\| \leq K\alpha^{t-\tau} \|\xi - \phi^*(\tau)\|$ for all $\tau \leq t$, $\xi \in B_\delta(\phi^*(\tau))$

and an *unstable* solution ϕ^* is not stable. These stability notions relate as follows:

$$UES \Rightarrow UAS \Rightarrow US$$
$$\Downarrow \qquad \Downarrow \qquad \Downarrow$$
$$ES \Rightarrow AS \Rightarrow S$$

and for periodic solutions ϕ^* to periodic equations (A.1), stability, respectively, asymptotic stability, is always uniform. Stability properties of the solution ϕ^* to equation (A.1) carry over to the trivial solution to the *equations of perturbed motion*

$$\dot{x} = f(t, \phi^*(t) + x) \quad \text{or} \quad x_{t+1} = f(t, \phi^*(t) + x_t) \tag{A.2}$$

and vice versa. Finally, one denotes a solution ϕ^* to a one-dimensional equation (A.1) as *semi-stable*, if it attracts solutions starting nearby in $\xi \geq \phi^*(\tau)$ and repels those with $\phi^*(\tau) < \xi$ (or conversely).

We now suppose that the partial derivative $D_2 f : \mathcal{D} \to \mathbb{R}^{d \times d}$ exists as continuous function and consider the variational equations

$$\dot{x} = D_2 f(t, \phi^*(t))x \quad \text{or} \quad x_{t+1} = D_2 f(t, \phi^*(t))x_t. \tag{A.3}$$

Then our first main result on stability is as follows:

Theorem A.1.1 (Linearised Uniform Exponential Stability) *If a solution* ϕ^* : $\mathbb{I} \to \mathbb{R}^d$ *of equation* (A.1) *on an interval* $\mathbb{I}$ *bounded below satisfies*

$$\sup_{t \in \mathbb{I}} \|D_2 f(t, \phi^*(t))\| < \infty,$$
$$\lim_{x \to 0} \sup_{t \in \mathbb{I}} \|D_2 f(t, \phi^*(t) + x) - D_2 f(t, \phi^*(t))\| = 0, \tag{A.4}$$

then the following holds:

(a) $\Sigma(\phi^*) \subseteq \begin{cases} (-\infty, 0), & \mathbb{T} = \mathbb{R}, \\ (0,1), & \mathbb{T} = \mathbb{Z} \end{cases}$ *if and only if* ϕ^* *is uniformly exponentially stable (on* $\mathbb{I}$*).*

(b) If $\begin{cases} (0,\infty), & \mathbb{T} = \mathbb{R}, \\ (1,\infty), & \mathbb{T} = \mathbb{Z} \end{cases}$ *contains a spectral interval of* $\Sigma(\phi^*)$, *then* ϕ^* *is unstable.*

The subsequent example illustrates that the above boundedness assumption on the coefficients of (A.3) is not technical:

Example A.1.2 *The linear, planar difference equation*

$$x_{t+1} = \begin{pmatrix} |t| + 1 & 0 \\ 0 & \frac{1}{2} \end{pmatrix} x_t \tag{A.5}$$

has the dichotomy spectrum $\{\frac{1}{2}\} \subset (0,1)$. *Its transition matrix*

$$\Phi(t,s) = \begin{pmatrix} \prod_{r=s}^{t-1}(|r| + 1) & 0 \\ 0 & 2^{s-t} \end{pmatrix} \quad \text{for all } s \le t$$

shows that unbounded solutions to (A.5) *exist. Therefore,* (A.5) *is unstable.*

On the entire line, that is $\mathbb{I} = \mathbb{R}$ or $\mathbb{I} = \mathbb{Z}$, in assertion (a) the condition on the spectrum $\Sigma(\phi^*)$ is still sufficient for uniform exponential stability (on $\mathbb{I}$). Moreover, then assertion (b) follows from the existence of an unstable manifold (fibre bundle) associated with ϕ^*. However, our subsequent argument merely assumes (A.3) to be given on a half line unbounded above.

Proof. The proof is essentially based on the Variation of Constants formula and the Grönwall lemma. Since both tools are standard in discrete and continuous time, we only present the case $\mathbb{T} = \mathbb{R}$ and refer to [186, p. 100, Theorem 3.1.16(a), respectively, p. 348, Proposition A.2.1(a)] for the appropriate discrete time results.

First, we restrict to properties for the trivial solution to equation (A.2), which is defined in an R-neighbourhood of 0 uniformly in $t \in \mathbb{I}$. Let $\tau \in \mathbb{I}$ and suppose $\hat{\varphi}$ stands for the general solution to (A.2). Note that $\hat{\varphi}(\cdot, \tau, x_\tau)$ is also a solution to the linearly inhomogeneous equation $\dot{x} = D_2 f(t, \phi^*(t))x + F(t, \hat{\varphi}(t, \tau, x_\tau))$ with

$$F(t,x) := f(t, \phi^*(t) + x) - f(t, \phi^*(t)) - D_2 f(t, \phi^*(t))x$$

and the Variation of Constants formula (see [96, p. 81, Theorem. 1.1]) yields

$$\hat{\varphi}(t,\tau,x_\tau) = \Phi(t,\tau)x_\tau + \int_\tau^t \Phi(t,s)F(s, \hat{\varphi}(s,\tau,x_\tau))\,ds \tag{A.6}$$

as long as $\hat{\varphi}(\cdot, \tau, x_\tau)$ exists. Second, by (A.4), for all $M > 0$, there is a $\rho > 0$ with

$$\|F(t,x)\| \le M\,\|x\| \quad \text{for all } t \in \mathbb{I},\ x \in \bar{B}_\rho(0). \tag{A.7}$$

(a) ($\Rightarrow$) Due to $\Sigma(\phi^*) \subseteq (-\infty, 0)$, there exist reals $K \ge 1$ and $\alpha < 0$ such that

$$\|\Phi(t,\tau)\| \le Ke^{\alpha(t-\tau)} \quad \text{for all } \tau \le t. \tag{A.8}$$

We choose $M \in (0, -\frac{\alpha}{K})$ and obtain $\alpha + KM < 0$. Finally, given an initial value $x_\tau \in B_\rho(0)$, we set $T^*(x_\tau) := \sup \{T \geq \tau : \|\hat{\varphi}(t, \tau, x_\tau)\| \leq \rho$ for all $\tau \leq t \leq T\}$ as exit time at which $\hat{\varphi}(\cdot, \tau, x_\tau)$ leaves the ρ-ball around the zero solution for the first time. This definition includes $T^*(x_\tau) = \infty$, when the solution stays in $B_\rho(0)$.

(I) We show that every initial value $x_\tau \in B_\rho(0)$ yields an estimate

$$\|\hat{\varphi}(t, \tau, x_\tau)\| \leq K e^{(\alpha + KM)(t-\tau)} \|x_\tau\| \quad \text{for all } \tau \leq t \leq T^*(x_\tau). \tag{A.9}$$

Thereto, on this interval $[\tau, T^*(x_\tau)]$, (A.6) leads to the estimate

$$\|\hat{\varphi}(t, \tau, x_\tau)\| \overset{(A.8)}{\leq} K e^{\alpha(t-\tau)} \|x_\tau\| + K \int_\tau^t e^{\alpha(t-s)} \|F(s, \hat{\varphi}(s, \tau, x_\tau))\| \, ds$$

$$\overset{(A.7)}{\leq} K e^{\alpha(t-\tau)} \|x_\tau\| + KM \int_\tau^t e^{\alpha(t-s)} \|\hat{\varphi}(s, \tau, x_\tau)\| \, ds$$

and multiplication with $e^{\alpha(\tau-t)}$ implies the inequality

$$e^{\alpha(\tau-t)} \|\hat{\varphi}(t, \tau, x_\tau)\| \leq K \|x_\tau\| + KM \int_\tau^t e^{\alpha(\tau-s)} \|\hat{\varphi}(s, \tau, x_\tau)\| \, ds$$

for all $\tau \leq t \leq T^*(x_\tau)$. This relation combined with the Grönwall lemma [96, p. 36, Corollary 6.6] yields

$$e^{\alpha(\tau-t)} \|\hat{\varphi}(t, \tau, x_\tau)\| \leq K e^{KM(t-\tau)} \|x_\tau\| \quad \text{for all } \tau \leq t \leq T^*(x_\tau),$$

which is obviously equivalent to (A.9).

(II) In order to conclude the proof for sufficiency, let us exploit (A.9) where $e^{(\alpha + KM)(t-\tau)} \in [0, 1)$ is strictly decreasing in $t \geq \tau$. First, given initial values $x_\tau \in B_{\rho/K}(0)$, we obtain that $\|\hat{\varphi}(t, \tau, x_\tau)\| \leq K e^{(\alpha + KM)(t-\tau)} \rho \leq \rho$ and hence $T^*(x_\tau) = \infty$. Thus, (A.9) holds for all $t \geq \tau$ and the trivial solution to (A.2) is uniformly exponentially stable.

($\Leftarrow$) By assumption, there exist reals $K \geq 1$, $\alpha < 0$ and $\delta > 0$ with

$$\|\hat{\varphi}(t, \tau, x_\tau)\| \leq K e^{\alpha(t-\tau)} \|x_\tau\| \quad \text{for all } \tau \leq t, \ x_\tau \in B_\delta(0) \tag{A.10}$$

and we mimic the discrete time case studied in [95]. For fixed $\beta \in (\alpha, 0)$, we choose $M \in (0, \frac{\beta - \alpha}{K})$, and using (A.4) there exists a $\rho > 0$ such that (A.7) holds. Provided $\rho_0 \in (0, \min\{\delta, \frac{\rho}{K}\})$, then (A.10) implies $\|\hat{\varphi}(t, \tau, x_\tau)\| \leq K \|x_\tau\| \leq K \rho_0 < \rho$ for every $\tau \leq t$, $x_\tau \in B_{\rho_0}(0)$ and

$$\|\Phi(t, \tau)x_\tau\| \overset{(A.6)}{\leq} \|\hat{\varphi}(t, \tau, x_\tau)\| + M \int_\tau^t \|\Phi(t, s)\| \|\hat{\varphi}(s, \tau, x_\tau)\| \, ds$$

$$\overset{(A.10)}{\leq} K e^{\alpha(t-\tau)} \rho_0 + MK\rho_0 \int_\tau^t e^{\alpha(s-\tau)} \|\Phi(t, s)\| \, ds.$$

Because of $\alpha < \beta$, this guarantees

$$e^{\beta(\tau-t)}\|\Phi(t,\tau)x_\tau\| \le K\rho_0 + MK\rho_0 \int_\tau^t e^{(\alpha-\beta)(s-\tau)}e^{\beta(s-t)}\|\Phi(t,s)\|\,\mathrm{d}s,$$

with $\omega(t) := \max_{s\in[\tau,t]} e^{\beta(s-t)}\|\Phi(t,s)\|$ results

$$e^{\beta(\tau-t)}\|\Phi(t,\tau)x_\tau\| \le K\rho_0 + MK\rho_0\omega(t)\int_\tau^t e^{(\alpha-\beta)(s-\tau)}\,\mathrm{d}s \le K\rho_0 + \frac{MK\rho_0}{\beta-\alpha}\omega(t)$$

and consequently

$$e^{\beta(\tau-t)}\|\Phi(t,\tau)\| = e^{\beta(\tau-t)}\sup_{\|x\|\le 1}\|\Phi(t,\tau)x\| = \frac{1}{\rho_0}e^{\beta(\tau-t)}\sup_{\|x\|\le\rho_0}\|\Phi(t,\tau)x\|$$
$$\le K + \frac{KM}{\beta-\alpha}\omega(t)\quad\text{for all }\tau\le t.$$

Therefore, $\omega(t) \le K + \frac{MK}{\beta-\alpha}\omega(t)$, and due to $\frac{MK}{\beta-\alpha} < 1$ by the choice of M, this shows $e^{\beta(\tau-t)}\|\Phi(t,\tau)\| \le \frac{K(\beta-\alpha)}{\beta-\alpha-MK}$. Thanks to $\beta < 0$, we see that (A.3) has an exponential dichotomy with projector $P(t) \equiv \mathrm{id}$ on $\mathbb{I}$ and thus $\Sigma(\phi^*) \subseteq (-\infty,0)$.

(b) Let σ denote a spectral interval in $(0,\infty)$. We start with the substitution $w(t) := e^{-\gamma t}x(t)$ for some $\gamma \in (\mathbb{R}\setminus\Sigma(\phi^*))\cap[0,\min\sigma)$ yielding the equation

$$\dot{w} = \left(D_2 f(t,\phi^*(t)) - \gamma\,\mathrm{id}\right)w + \tfrac{1}{e^{\gamma t}}F(t,e^{\gamma t}w),$$

whose general solution ψ is related to the general solution $\hat\varphi$ of (A.2) by

$$\psi(t,\tau,w_\tau) = e^{-\gamma t}\hat\varphi(t,\tau,x_\tau),\qquad w_\tau = e^{-\gamma\tau}x_\tau. \tag{A.11}$$

Because of $\min\sigma > 0$, the shifted variational equation

$$\dot{w} = (D_2 f(t,\phi^*(t)) - \gamma\,\mathrm{id})w \tag{A.12}$$

has an exponential dichotomy on $\mathbb{I}$ with projectors $P(t) \neq \mathrm{id}$ for all $t\in\mathbb{I}$, i.e. there exist reals $K \ge 1$, $\alpha < 0$ such that

$$\|\Phi_\gamma(t,s)P(s)\| \le Ke^{\alpha(t-s)},\qquad \|\Phi_\gamma(s,t)[\mathrm{id}-P(t)]\| \le Ke^{\alpha(t-s)} \tag{A.13}$$

for all $s,t\in\mathbb{I}$, $s\le t$, where Φ_γ is the transition matrix to (A.12). Hence, we obtain

$$\int_\tau^t \|\Phi_\gamma(t,s)P(s)\|\,\mathrm{d}s + \int_t^\infty \|\Phi_\gamma(t,s)[\mathrm{id}-P(s)]\|\,\mathrm{d}s$$
$$\overset{(A.13)}{\le} K\int_\tau^t e^{\alpha(t-s)}\,\mathrm{d}s + K\int_t^\infty e^{\alpha(s-t)}\,\mathrm{d}s \le -\frac{2K}{\alpha} =: C \tag{A.14}$$

for all $\tau\le t$ and choosing $M\in(0,C^{-1})$ in (A.7) implies $1 - CM > 0$.

(I) We show that if $\hat\varphi(\cdot,\tau,x_\tau)$ satisfies

$$\|\hat{\varphi}(t,\tau,x_\tau)\| \le \rho \quad \text{for all } \tau \le t, \tag{A.15}$$

then

$$\|\hat{\varphi}(t,\tau,x_\tau)\| \le e^{\gamma(t-\tau)}(1-CM)^{-1}K \,\|P(\tau)x_\tau\| \quad \text{for all } \tau \le t. \tag{A.16}$$

Indeed, the assumption (A.15) yields

$$\left\|\tfrac{1}{e^{\gamma t}}F(t,\hat{\varphi}(t,\tau,x_\tau))\right\| \overset{(A.7)}{\le} \tfrac{M}{e^{\gamma t}}\|\hat{\varphi}(t,\tau,x_\tau)\| \le \tfrac{M}{e^{\gamma\tau}}\rho \quad \text{for all } \tau \le t.$$

Combined with (A.14), this guarantees that

$$\eta(t) := \psi(t,\tau,w_\tau) - \Phi_\gamma(t,\tau)P(\tau)w_\tau + \int_\tau^t \frac{1}{e^{\gamma s}}\Phi_\gamma(t,s)P(s)F(s,\hat{\varphi}(s,\tau,x_\tau))\,ds$$
$$- \int_t^\infty \frac{1}{e^{\gamma s}}\Phi_\gamma(t,s)[\text{id}-P(s)]F(s,\hat{\varphi}(s,\tau,x_\tau))\,ds$$

defines a bounded function $\eta : [\tau,\infty) \to \mathbb{R}^d$. Moreover, the curious reader might verify that η indeed solves (A.12) and satisfies $P(\tau)\eta(\tau) = 0$. Since (A.12) admits an exponential dichotomy on $[\tau,\infty)$, this is only possible if $\eta(t) \equiv 0$ on $[\tau,\infty)$ (cf. [54, p. 38]). Thus, we obtain

$$\psi(t,\tau,w_\tau) = \Phi_\gamma(t,\tau)P(\tau)w_\tau - \int_\tau^t \frac{1}{e^{\gamma s}}\Phi_\gamma(t,s)P(s)F(s,\hat{\varphi}(s,\tau,x_\tau))\,ds$$
$$+ \int_t^\infty \frac{1}{e^{\gamma s}}\Phi_\gamma(t,s)[\text{id}-P(s)]F(s,\hat{\varphi}(s,\tau,x_\tau))\,ds,$$

according to (A.7) this implies

$$\|\psi(t,\tau,w_\tau)\| \le Ke^{\alpha(t-\tau)}\|P(\tau)w_\tau\| + M\int_\tau^t \|\Phi_\gamma(t,s)P(s)\|\frac{\|\hat{\varphi}(s,\tau,x_\tau)\|}{e^{\gamma s}}\,ds$$
$$+ M\int_t^\infty \|\Phi_\gamma(t,s)[\text{id}-P(s)]\|\frac{\|\hat{\varphi}(s,\tau,x_\tau)\|}{e^{\gamma s}}\,ds,$$

a combination of (A.11), (A.14) and $\gamma \ge 0$ gives

$$\|\psi(t,\tau,w_\tau)\| \le Ke^{\alpha(t-\tau)}\|P(\tau)w_\tau\| + CM\sup_{\tau \le s}\|\psi(s,\tau,w_\tau)\| \quad \text{for all } \tau \le t$$

and finally the estimate $\|\psi(t,\tau,w_\tau)\| \le e^{-\gamma\tau}(1-CM)^{-1}K\,\|P(\tau)x_\tau\|$ holds. Thanks to (A.11), we arrive at (A.16).

(II) Suppose that the trivial solution to (A.2) is stable, i.e. for $\varepsilon > 0$ there exists a $\delta > 0$ such that $x_\tau \in B_\delta(0)$ implies $\|\hat{\varphi}(t,\tau,x_\tau)\| < \varepsilon$ for all $\tau \le t$. Since $P(\tau) \neq \text{id}$ holds, one can choose a nonzero initial value x_τ in $N(P(\tau))$ and (A.16) implies $0 < \|x_\tau\| = \|\hat{\varphi}(\tau,\tau,x_\tau)\| = 0$, which is a contradiction. $\square$

The authors are not aware of a related characterisation in terms of the Lyapunov spectrum. Nevertheless a sufficient condition can be given based on the following: let $\Sigma_{\text{lyap}}(\phi^*) = \{\chi_1, \dots, \chi_n\}$ denote the Lyapunov spectrum of the variational equations (A.3) and suppose that a Lyapunov exponent $\chi_j \in \mathbb{R}$ has the multiplicity $\kappa_j \in \mathbb{N}$. Given $\tau \in \mathbb{I}$, we define the *regularity coefficient*

$$
\Gamma(\phi^*) := \begin{cases} \sum_{j=1}^n \kappa_j \chi_j - \liminf_{t \to \infty} \frac{1}{t} \int_\tau^t \operatorname{tr} D_2 f(s, \phi^*(s)) \, ds, & \mathbb{T} = \mathbb{R}, \\[2mm] \dfrac{\prod_{j=1}^n \chi_j^{\kappa_j}}{\liminf_{t \to \infty} \sqrt[t]{\prod_{s=\tau}^{t-1} |\det D_2 f(s, \phi^*(s))|}}, & \mathbb{T} = \mathbb{Z} \end{cases}
$$

Of a solution $\phi^* : \mathbb{I} \to \mathbb{R}^d$ to (A.1). We speak of a *regular solution*, if $\Gamma(\phi^*) = 0$ (in continuous time $\mathbb{T} = \mathbb{R}$) and, respectively, $\Gamma(\phi^*) = 1$ (in discrete time $\mathbb{T} = \mathbb{Z}$).

Theorem A.1.3 (Linearised Exponential Stability) *Let $\phi^* : \mathbb{I} \to \mathbb{R}^d$ be a solution of equation* (A.1) *satisfying*

$$
\sup_{t \in \mathbb{I}} \| D_2 f(t, \phi^*(t)) \| < \infty
$$

and that there exists a neighbourhood $U \subseteq \mathbb{R}^d$ of 0, $L \geq 0$, $\nu > 0$ with

$$
\| D_2 f(t, \phi^*(t) + x) - D_2 f(t, \phi^*(t)) \| \leq L \|x\|^\nu \quad \text{for all } t \in \mathbb{I} \tag{A.17}
$$

and $x \in U$. If one of the conditions

(a) $\mathbb{T} = \mathbb{R}$ *and* $\max \{ \max \Sigma_{lyap}(\phi^*), \nu \max \Sigma_{lyap}(\phi^*) \} + \Gamma(\phi^*) < 0$
(b) $\mathbb{T} = \mathbb{Z}$, $D_2 f(t, \phi^*(t))$, $t \in \mathbb{I}$, *is invertible and*

$$
\max \{ \max \Sigma_{lyap}(\phi^*), \Gamma(\phi^*)(\max \Sigma_{lyap}(\phi^*))^\nu \} < 1
$$

holds, then ϕ^* is exponentially stable.

Hence, the stability of regular solutions is given by the Lyapunov spectrum alone.

Proof. (a) See [24, p. 71, Theorem 3.9], where the explicit form of the regularity coefficient $\Gamma(\phi^*)$ is due to [153, p. 37, Definition 3.6].
    (b) Let us begin with preparations, set $\chi := \max \Sigma_{\text{lyap}}(\phi^*) < 1$ and choose $\tau \in \mathbb{I}$ fixed. We write Eq. (A.2) of perturbed motion in semilinear form

$$
x_{t+1} = D_2 f(t, \phi^*(t)) x_t + F(t, x_t)
$$

with the nonlinearity $F(t, x) := f(t, \phi^*(t) + x) - f(t, \phi^*(t)) - D_2 f(t, \phi^*(t)) x$ and the Mean Value Theorem implies

$$
F(t, x) - F(t, \bar{x}) = f(t, \phi^*(t) + x) - f(t, \phi^*(t) + \bar{x}) - D_2 f(t, \phi^*(t))[x - \bar{x}]
$$
$$
= \int_0^1 [D_2 f(t, \phi^*(t) + (1 - h)\bar{x} + hx) - D_2 f(t, \phi^*(t))] \, dh[x - \bar{x}].
$$

Choose $\rho > 0$ so small that $\bar{B}_\rho(0) \subseteq U$ holds. If $x, \bar{x} \in \bar{B}_\rho(0)$, then the convexity of $\bar{B}_\rho(0)$ yields $(1 - h)\bar{x} + hx \in \bar{B}_\rho(0)$ for all $h \in [0, 1]$ and therefore

$$\|F(t, x) - F(t, \bar{x})\| \tag{A.18}$$

$$\leq \int_0^1 \|D_2 f(t, \phi^*(t) + (1 - h)\bar{x} + hx) - D_2 f(t, \phi^*(t))\| \, dh \, \|x - \bar{x}\|$$

$$\overset{(A.17)}{\leq} L \int_0^1 \|(1 - h)\bar{x} + hx\|^\nu \, dh \, \|x - \bar{x}\| \leq L \left(\|x\| + \|\bar{x}\|\right)^\nu \|x - \bar{x}\|$$

holds for all $\tau \leq t$. Our assumption $\max\{\chi, \Gamma(\phi^*)\chi^\nu\} < 1$ guarantees that $\varepsilon > 0$ can be chosen so small such that also $\alpha := \chi + \varepsilon$ and $\gamma := \Gamma(\phi^*) + \varepsilon$ satisfy

$$\max\{\alpha, \gamma\alpha^\nu\} < 1. \tag{A.19}$$

This allows us to conclude from [147, p. 26, Theorem 2] that there is a $K \geq 1$ with

$$\|\Phi(t, s)\| \leq K\alpha^{t-s}\gamma^s \quad \text{for all } s \leq t. \tag{A.20}$$

(I) We equip the linear space $\ell_\alpha := \{\phi : \mathbb{Z}_\tau^+ \to \mathbb{R}^d | \sup_{\tau \leq t} \|\phi(t)\| \, \alpha^{\tau-t} < \infty\}$ with the norm $\|\phi\|_\alpha := \sup_{\tau \leq t} \|\phi(t)\| \, \alpha^{\tau-t}$, which leads to a Banach space. Given $\xi \in \mathbb{R}^d$ and $\phi \in \ell_\alpha^+$, consider the formally defined operators

$$S\xi := \Phi(\cdot, \tau)\xi, \qquad J(\phi) := \sum_{s=\tau}^{t-1} \Phi(\cdot, s+1)F(s, \phi(s)).$$

First, from (A.20), we deduce $\|[S\xi](t)\| \leq \|\Phi(t, \tau)\| \|\xi\| \leq K\alpha^{t-\tau}\gamma^\tau \|\xi\|$ for all $\tau \leq t$, which shows that $S : \mathbb{R}^d \to \ell_\alpha$ is well defined and satisfies

$$\|S\xi\|_\alpha \leq K\gamma^\tau \|\xi\|. \tag{A.21}$$

Second, the elementary inequality

$$\|\phi(t)\| \leq \|\phi\|_\alpha \, \alpha^{t-\tau} \overset{(A.19)}{\leq} \|\phi\|_\alpha \quad \text{for all } \tau \leq t, \phi \in \ell_\alpha$$

guarantees that sequences $\phi, \bar{\phi} \in \bar{B}_\rho(0) \subseteq \ell_\alpha$ satisfy $\phi(t), \bar{\phi}(t) \in U$ for all $t \geq \tau$. This allows us to derive

$$\|[J(\phi)](t) - [J(\bar{\phi})](t)\| \, \alpha^{\tau-t}$$

$$\leq \sum_{s=\tau}^{t-1} \|\Phi(t, s+1)\| \, \|F(s, \phi(s)) - F(s, \bar{\phi}(s))\| \, \alpha^{\tau-t}$$

$$\overset{(A.20)}{\leq} K\frac{\gamma}{\alpha} \sum_{s=\tau}^{t-1} \alpha^{\tau-s}\gamma^s \, \|F(s, \phi(s)) - F(s, \bar{\phi}(s))\|$$

$$\stackrel{(A.18)}{\leq} KL\frac{\gamma}{\alpha}\sum_{s=\tau}^{t-1}\alpha^{\tau-s}\gamma^{s}(\|\phi(s)\|+\|\bar{\phi}(s)\|)^{\nu}\|\phi(s)-\bar{\phi}(s)\|$$

$$\leq KL\frac{\gamma}{\alpha}(2\rho)^{\nu}\sum_{s=\tau}^{t-1}\alpha^{(\nu-1)(s-\tau)}\gamma^{s}\|\phi(s)-\bar{\phi}(s)\|$$

$$\leq KL\frac{\gamma}{\alpha}(2\rho)^{\nu}\sum_{s=\tau}^{t-1}\alpha^{\nu(s-\tau)}\gamma^{s}\|\phi-\bar{\phi}\|_{\alpha}$$

$$\stackrel{(A.19)}{\leq} KL\frac{\gamma^{\tau+1}}{\alpha-\gamma\alpha^{\nu+1}}(2\rho)^{\nu}\|\phi-\bar{\phi}\|_{\alpha}\quad\text{for all }\tau\leq t$$

and passing to the least upper bound over $\tau \leq t$ shows the Lipschitz condition

$$\|J(\phi)-J(\bar{\phi})\|_{\alpha}\leq KL\frac{\gamma^{\tau+1}}{\alpha-\gamma\alpha^{\nu+1}}(2\rho)^{\nu}\|\phi-\bar{\phi}\|_{\alpha}.\qquad(A.22)$$

Due to $J(0) = 0$, this implies that $J : \bar{B}_{\rho}(0) \to \ell_{\alpha}$ is well defined. Moreover, for sufficiently small $\rho > 0$, it is $L_{\tau} := KL\frac{\gamma^{\tau+1}}{\alpha-\gamma\alpha^{\nu+1}}(2\rho)^{\nu} < 1$. Hence,

$$\|S\xi+J(\phi)\|_{\alpha} \leq \|S\xi\|_{\alpha}+\|J(\phi)\|_{\alpha} \stackrel{(A.21)}{\leq} K\gamma^{\tau}\|\xi\|+\|J(\phi)-J(0)\|_{\alpha}$$
$$\stackrel{(A.22)}{\leq} K\gamma^{\tau}\|\xi\|+L_{\tau}\|\phi\|_{\alpha}\leq\rho\quad\text{for all }\xi\in B_{\delta}(0),\ \phi\in\bar{B}_{\rho}(0)$$

holds, if $\delta \leq \frac{(1-L_{\tau})\rho}{K\gamma^{\tau}}$. In conclusion, for $\xi \in B_{\delta}(0)$, the map $S\xi + J : \bar{B}_{\rho}(0) \to \bar{B}_{\rho}(0)$ is a contraction on the complete metric space $\bar{B}_{\rho}(0) \subseteq \ell_{\alpha}$. By the Contraction Mapping Principle, this yields that there exists a unique $\phi^{\star} \in \bar{B}_{\rho}(0) \subseteq \ell_{\alpha}$ satisfying $\phi^{\star} = S\xi + J(\phi^{\star})$.

(II) Due to the definition of the operators S and J, the fixed-point relation for $\phi^{\star} \in \ell_{\alpha}$ explicitly reads as

$$\phi^{\star}(t)=\Phi(t,\tau)\xi+\sum_{s=\tau}^{t-1}\Phi(t,s+1)F(s,\phi^{\star}(s))\quad\text{for all }\tau\leq t.$$

Due to the Variation of Constants formula [186, p. 100, Theorem 3.1.16(a)], this means $\phi^{\star}(t) = \hat{\varphi}(t,\tau,\xi)$ with the general solution $\hat{\varphi}$ to Eq. (A.2) of perturbed motion. Whence, if $\xi \in B_{\delta}(0)$ holds, then

$$\|\hat{\varphi}(t,\tau,\xi)\|=\|\phi^{\star}(t)\|\leq\|\phi^{\star}\|_{\alpha}\,\alpha^{t-\tau}\leq\frac{K\gamma^{\tau}\alpha^{-\tau}}{1-L_{\tau}}\alpha^{t}\|\xi\|\quad\text{for all }\tau\leq t$$

results. This establishes that the trivial solution of equation (A.2) and therefore also the solution ϕ^* of (A.1) are exponentially stable.

A.2 Bohl and Lyapunov Exponents

Lyapunov or *characteristic exponents* are a tool to measure the exponential growth of functions, which are typically solutions to differential or difference equations [1, 24, 147, 154]. In nonautonomous stability theory, however, it is beneficial to have additional information on the *uniform* exponential growth described by *Bohl exponents* [59, 101]. Of particular importance in numerical approximations of dynamical spectra [62–64] are such exponents in the special case of real functions.

Continuous Time

Let $\mathbb{I} \subseteq \mathbb{R}$ denote an unbounded interval and suppose $a : \mathbb{I} \to \mathbb{R}$ is locally integrable. The *upper Bohl exponent* of a is defined by

$$\overline{\beta}_{\mathbb{I}}(a) := \inf \left\{ \omega \in \mathbb{R} : \sup_{s \leq t,\, s,t \in \mathbb{I}} \int_s^t (a(r) - \omega)\, \mathrm{d}r < \infty \right\}$$

and the *lower Bohl exponent* by

$$\underline{\beta}_{\mathbb{I}}(a) := \sup \left\{ \omega \in \mathbb{R} : -\infty < \inf_{s \leq t,\, s,t \in \mathbb{I}} \int_s^t (a(r) - \omega)\, \mathrm{d}r \right\}.$$

This readily implies the linearity relations

$$\overline{\beta}_{\mathbb{I}}(\nu + \mu a) = \nu + \mu \overline{\beta}_{\mathbb{I}}(a), \quad \underline{\beta}_{\mathbb{I}}(\nu + \mu a) = \nu + \mu \underline{\beta}_{\mathbb{I}}(a) \quad \text{for all } \nu \in \mathbb{R},\, \mu > 0.$$

Finiteness of these exponents can be characterised via [59, p. 119, Theorem 4.2] as

$$\overline{\beta}_{\mathbb{I}}(a) < \infty \quad \Leftrightarrow \quad \sup_{\substack{0 \leq t-s \leq 1 \\ s,t \in \mathbb{I}}} \int_s^t a(r)\, \mathrm{d}r < \infty,$$

$$-\infty < \underline{\beta}_{\mathbb{I}}(a) \quad \Leftrightarrow \quad -\infty < \inf_{\substack{0 \leq t-s \leq 1 \\ s,t \in \mathbb{I}}} \int_s^t a(r)\, \mathrm{d}r,$$

and they are related by $-\beta(a) = \overline{\beta}(-a)$. In case $\underline{\beta}_{\mathbb{I}}(a) = \overline{\beta}_{\mathbb{I}}(a)$, one says the function a has a *strict Bohl exponent* $\beta_{\mathbb{I}}(a)$. On each subinterval $\mathbb{J} \subseteq \mathbb{I}$, the inequality

$$\underline{\beta}_{\mathbb{I}}(a) \leq \underline{\beta}_{\mathbb{J}}(a) \leq \overline{\beta}_{\mathbb{J}}(a) \leq \overline{\beta}_{\mathbb{I}}(a) \tag{A.23}$$

holds. From now on, let $\mathbb{I} = \mathbb{R}_\tau^+$ for some $\tau \in \mathbb{R}$. Then the characterizations

$$\overline{\beta}_{\mathbb{I}}(a) = \limsup_{s,T \to \infty} \frac{1}{T} \int_s^{s+T} a(r)\, \mathrm{d}r, \qquad \underline{\beta}_{\mathbb{I}}(a) = \liminf_{s,T \to \infty} \frac{1}{T} \int_s^{s+T} a(r)\, \mathrm{d}r$$

hold with $\beta_{\mathbb{I}}(a) = \lim_{s,T\to\infty} \frac{1}{T} \int_s^{s+T} a(r)\,dr$ for strict Bohl exponents.

One defines the *upper Lyapunov exponent* of a as

$$\overline{\chi}(a) := \inf\left\{\omega \in \mathbb{R} : \sup_{t\in\mathbb{R}_\tau^+} \int_\tau^t (a(r) - \omega)\,dr < \infty\right\}$$

and the *lower Lyapunov exponent* of a as

$$\underline{\chi}(a) := \sup\left\{\omega \in \mathbb{R} : -\infty < \inf_{t\in\mathbb{R}_\tau^+} \int_\tau^t (a(r) - \omega)\,dr\right\}$$

and obtains the relations

$$\overline{\chi}(\nu + \mu a) = \nu + \mu\overline{\chi}(a), \quad \underline{\chi}(\nu + \mu a) = \nu + \mu\underline{\chi}(a) \quad \text{for all } \nu \in \mathbb{R},\ \mu > 0.$$

Both Lyapunov exponents are independent of $\tau \in \mathbb{I}$, allow the characterisation

$$\overline{\chi}(a) = \limsup_{t\to\infty} \frac{1}{t-\tau} \int_\tau^t a(r)\,dr, \quad \underline{\chi}(a) = \liminf_{t\to\infty} \frac{1}{t-\tau} \int_\tau^t a(r)\,dr,$$

see [59, p. 118], and are related by $-\underline{\chi}(a) = \overline{\chi}(-a)$. For a *strict Lyapunov exponent* $\chi(a)$ of a, one has $\underline{\chi}(a) = \overline{\chi}(a)$ and consequently $\chi(a) = \lim_{t\to\infty} \frac{1}{t-\tau} \int_\tau^t a(r)\,dr$.

Finally, Bohl and Lyapunov exponents are related by

$$\underline{\beta}_{\mathbb{I}}(a) \leq \underline{\chi}(a) \leq \overline{\chi}(a) \leq \overline{\beta}_{\mathbb{I}}(a), \tag{A.24}$$

and hence strict Bohl exponents imply strict Lyapunov exponents.

Example A.2.1 *For the functions* $a_i : \mathbb{R} \to \mathbb{R}$,

$$a_0(t) :\equiv \alpha \in \mathbb{R}, \quad a_1(t) = \sin t, \quad a_2(t) := \arctan(-t), \quad a_3(t) = \sin(\ln(1 + |t|)),$$

a direct computation yields the exponents

i	$\underline{\beta}_{\mathbb{R}}(a_i)$	$\overline{\beta}_{\mathbb{R}}(a_i)$	$\underline{\beta}_{\mathbb{R}_0^+}(a_i)$	$\overline{\beta}_{\mathbb{R}_0^+}(a_i)$	$\underline{\beta}_{\mathbb{R}_0^-}(a_i)$	$\overline{\beta}_{\mathbb{R}_0^-}(a_i)$	$\overline{\chi}(a)$
0	α	α	α	α	α	α	α
1	0	0	0	0	0	0	0
2	$-\pi/2$	$\pi/2$	$-\pi/2$	$-\pi/2$	$\pi/2$	$\pi/2$	$-\pi/2$
3	-1	1	-1	1	-1	1	$\sqrt{2}/2$

Generalising the asymptotically constant example a_2 *to general continuous functions* $a : \mathbb{R} \to \mathbb{R}$ *satisfying* $\lim_{t\to\pm\infty} a(t) = \alpha_\pm \in \mathbb{R}$ *gives*

$$\overline{\beta}_{\mathbb{R}}(a) = \max\{\alpha_-, \alpha_+\}, \qquad \underline{\beta}_{\mathbb{R}}(a) = \min\{\alpha_-, \alpha_+\},$$

$$\overline{\beta}_{\mathbb{R}_0^+}(a) = \underline{\beta}_{\mathbb{R}_0^+}(a) = \alpha_+, \qquad \overline{\beta}_{\mathbb{R}_0^-}(a) = \underline{\beta}_{\mathbb{R}_0^-}(a) = \alpha_-.$$

An example illustrating the second inequality in (A.24) is the following:

Example A.2.2 (Perron) *For the function* $a : (0, \infty) \to \mathbb{R}$,

$$a(t) := \sin(\ln(t)) + \cos(\ln(t)),$$

one has $\int_\tau^t a(r)\, dr = t\sin(\ln(t)) - \tau\sin(\ln(\tau))$ *and thus* $\overline{\chi}(a) \leq 1$. *On the other hand, it is shown in [101, p. 259, Example 3.3.12] that* $\overline{\beta}_\mathbb{I}(a) = \sqrt{2}$.

Discrete Time

The discrete time situation of an unbounded discrete interval $\mathbb{I} \subseteq \mathbb{Z}$ and a real sequence $a = (a_t)_{t \in \mathbb{I}}$ is largely analogous to the above continuous time case. Now the *upper Bohl exponent* of a is defined by

$$\overline{\beta}_\mathbb{I}(a) := \inf \left\{ \omega > 0 : \sup_{s \leq t,\, s, t \in \mathbb{I}} \prod_{r=s}^{t-1} \frac{|a_r|}{\omega} < \infty \right\}$$

and the *lower Bohl exponent* by

$$\underline{\beta}_\mathbb{I}(a) := \sup \left\{ \omega > 0 : 0 < \inf_{s \leq t,\, s, t \in \mathbb{I}} \prod_{r=s}^{t-1} \frac{|a_r|}{\omega} \right\}.$$

This implies the positive homogeneity

$$\overline{\beta}_\mathbb{I}(\nu a) = |\nu|\, \overline{\beta}_\mathbb{I}(a), \qquad \underline{\beta}_\mathbb{I}(\nu a) = |\nu|\, \underline{\beta}_\mathbb{I}(a) \quad \text{for all } \nu \in \mathbb{R},$$

while finiteness of these exponents is guaranteed by the following conditions:

$$\sup_{t \in \mathbb{I}} |a_t| < \infty \quad \Rightarrow \quad \overline{\beta}_\mathbb{I}(a) < \infty,$$

$$0 < \inf_{t \in \mathbb{I}} |a_t| \quad \Rightarrow \quad 0 < \underline{\beta}_\mathbb{I}(a)$$

and they are related by $\underline{\beta}(a)^{-1} = \overline{\beta}(\frac{1}{a})$. On discrete subintervals $\mathbb{J} \subseteq \mathbb{I}$, the inequality (A.23) persists. If $\underline{\beta}_\mathbb{I}(a) = \overline{\beta}_\mathbb{I}(a)$, then the sequence a is said to have a *strict Bohl exponent* $\beta_\mathbb{I}(a)$.

From now we restrict to discrete intervals $\mathbb{I} = \mathbb{Z}_\tau^+$ with some $\tau \in \mathbb{Z}$. This yields the characterizations

$$\overline{\beta}_\mathbb{I}(a) = \limsup_{s, T \to \infty} \sqrt[T]{\prod_{r=s}^{s+T-1} |a_r|}, \qquad \underline{\beta}_\mathbb{I}(a) = \liminf_{s, T \to \infty} \sqrt[T]{\prod_{r=s}^{s+T-1} |a_r|}$$

with $\beta_\mathbb{I}(a) = \lim_{s, T \to \infty} \sqrt[T]{\prod_{r=s}^{s+T-1} |a_r|}$ for strict Bohl exponents.

One introduces the *upper Lyapunov exponent* of a as

$$\overline{\chi}(a) := \inf\left\{\omega > 0 : \sup_{t \in \mathbb{Z}_\tau^+} \prod_{r=s}^{t-1} \frac{|a_r|}{\omega} < \infty\right\}$$

and the *lower Lyapunov exponent* of a as

$$\underline{\chi}(a) := \sup\left\{\omega > 0 : 0 < \inf_{t \in \mathbb{Z}_\tau^+} \prod_{r=s}^{t-1} \frac{|a_r|}{\omega}\right\},$$

which leads to the relations

$$\overline{\chi}(\nu a) = |\nu|\,\overline{\chi}(a), \qquad\qquad \underline{\chi}(\nu a) = |\nu|\,\underline{\chi}(a) \quad \text{for all } \nu \in \mathbb{R}.$$

Both Lyapunov exponents are independent of $\tau \in \mathbb{I}$ and can be characterized as

$$\overline{\chi}(a) = \limsup_{T \to \infty} \sqrt[\tau+T]{\prod_{r=\tau}^{\tau+T-1} |a_r|}, \qquad \underline{\chi}(a) = \liminf_{T \to \infty} \sqrt[\tau+T]{\prod_{r=\tau}^{\tau+T-1} |a_r|}$$

and satisfy $\frac{1}{\underline{\chi}(a)} = \overline{\chi}(\frac{1}{a})$. A *strict Lyapunov exponent* $\chi(a)$ of a is determined via the condition $\underline{\chi}(a) = \overline{\chi}(a)$ and hence $\chi(a) = \lim_{T\to\infty} \sqrt[\tau+T]{\prod_{r=\tau}^{\tau+T-1} |a_r|}$ in this case. After all, (A.24) relates Bohl and Lyapunov exponents.

We denote the family of discrete subintervals of $\mathbb{I}$ with k elements by $\mathfrak{J}_k(\mathbb{I})$, i.e.,

$$\mathfrak{J}_k(\mathbb{I}) := \{\mathbb{J} \subseteq \mathbb{I} : \mathbb{J} \text{ is a discrete interval with } \#\mathbb{J} = k\}.$$

The *upper Bohl exponent* of a real sequence $(a_t)_{t \in \mathbb{I}}$ is given by

$$\overline{\beta}_{\mathbb{I}}(a) := \limsup_{k \to \infty} \sup_{\mathbb{J} \in \mathfrak{J}_k(\mathbb{I})} \sqrt[k]{\left|\prod_{s \in \mathbb{J}} a_s\right|}$$

and the *lower Bohl exponent* of $(a_t)_{t \in \mathbb{I}}$ reads as

$$\underline{\beta}_{\mathbb{I}}(a) := \liminf_{k \to \infty} \inf_{\mathbb{J} \in \mathfrak{J}_k(\mathbb{I})} \sqrt[k]{\left|\prod_{s \in \mathbb{J}} a_s\right|}.$$

On every discrete subinterval $\mathbb{J} \subseteq \mathbb{I}$, the Bohl exponents fulfil

$$0 \leq \inf_{t \in \mathbb{I}} |a_t| \leq \underline{\beta}_{\mathbb{I}}(a) \leq \underline{\beta}_{\mathbb{J}}(a) \leq \overline{\beta}_{\mathbb{J}}(a) \leq \overline{\beta}_{\mathbb{I}}(a) \leq \sup_{t \in \mathbb{I}} |a_t|. \tag{A.25}$$

Suppose now that $\mathbb{I}$ is unbounded above and let $\tau \in \mathbb{I}$. The *lower*, respectively, the *upper Lyapunov exponent*, of A is given by

$$\underline{\chi}(a) := \liminf_{t \to \infty} \sqrt[t]{\left|\prod_{s=\tau}^{t-1} a_s\right|}, \qquad\qquad \overline{\chi}(a) := \limsup_{t \to \infty} \sqrt[t]{\left|\prod_{s=\tau}^{t-1} a_s\right|}.$$

Lyapunov exponents are independent of $\tau \in \mathbb{I}$, as long as the a_t do not vanish. We denote them as *strict*, if $\underline{\chi}(a) = \overline{\chi}(a)$ holds. Furthermore, the exponents are positively homogeneous, that is

$$\overline{\beta}_{\mathbb{I}}(\nu a) = |\nu|\,\overline{\beta}_{\mathbb{I}}(a), \quad \underline{\beta}_{\mathbb{I}}(\nu a) = |\nu|\,\underline{\beta}_{\mathbb{I}}(a), \quad \overline{\chi}(\nu a) = |\nu|\,\overline{\chi}(a) \quad \text{for all } \nu \in \mathbb{R}.$$

In comparison to Bohl exponents, Lyapunov exponents satisfy

$$\underline{\beta}_{\mathbb{Z}}(a) \leq \underline{\beta}_{\mathbb{Z}_\tau^+}(a) \leq \underline{\chi}(a) \leq \overline{\chi}(a) \leq \overline{\beta}_{\mathbb{Z}_\tau^+}(a) \leq \overline{\beta}_{\mathbb{Z}}(a) \quad \text{for all } \tau \in \mathbb{Z}.$$

If $\mathbb{I}$ is unbounded above, then the above definition simplifies to

$$\overline{\beta}_{\mathbb{I}}(a) = \limsup_{k \to \infty} \sup_{\tau \in \mathbb{I}} \sqrt[k]{\left| \prod_{s=\tau}^{k+\tau-1} a_s \right|}, \qquad \underline{\beta}_{\mathbb{I}}(a) = \liminf_{k \to \infty} \inf_{\tau \in \mathbb{I}} \sqrt[k]{\left| \prod_{s=\tau}^{k+\tau-1} a_s \right|}.$$

References

1. L. Ya. Adrianova, *Introduction to linear systems of differential equations*, Translations of Mathematical Monographs, vol. 146, AMS, Providence, RI, 1995.
2. H.M. Alkhayuon, *Rate-induced transitions for parameter shift systems*, Ph.D. thesis, University of Exeter, 2018.
3. H.M. Alkhayuon and P. Ashwin, *Rate-induced tipping from periodic attractors: Partial tipping and connecting orbits*, Chaos **28** (2018), no. 3, 033608, 11.
4. A.I. Alonso, J. Hong, and R. Obaya, *Exponential dichotomy and trichotomy for difference equations*, Comput. Math. Appl. **38** (1999), no. 1, 41–49.
5. V. Anagnostopoulou and T. Jäger, *Nonautonomous saddle-node bifurcations: Random and deterministic forcing*, J. Differ. Equations **253** (2012), no. 2, 379–399.
6. V. Anagnostopoulou, T. Jäger, and G. Keller, *A model for the nonautonomous Hopf bifurcation*, Nonlinearity **28** (2015), no. 7, 2587–2616.
7. L. Arnold, *Random dynamical systems*, Monographs in Mathematics, Springer, Berlin etc., 1998.
8. L. Arnold, E. Oeljeklaus, and E. Pardoux, *Almost sure and moment stability for linear Ito equations*, Lyapunov exponents. Proceedings, Bremen 1984 (L. Arnold and V. Wihstutz, eds.), Springer, Berlin etc., 1986, pp. 129–159.
9. L. Arnold, N. Sri Namachchivaya, and K.R. Schenk-Hoppé, *Toward an understanding of stochastic Hopf bifurcation: A case study*, Int. J. Bifurcation Chaos **6** (1996), no. 11, 1947–1975.
10. L. Arnold and K. Xu, *Invariant measures for random dynamical systems, and a necessary condition for stochastic bifurcation from a fixed point*, Random & Computational Dynamics **2** (1994), no. 2, 165–182.
11. D.K. Arrowsmith and C.M. Place, *An introduction to dynamical systems.*, Cambridge University Press, Cambridge, 1990.

© The Author(s), under exclusive license to Springer Nature Switzerland AG 2023   139
V. Anagnostopoulou et al., *Nonautonomous Bifurcation Theory*, Frontiers
in Applied Dynamical Systems: Reviews and Tutorials 10,
https://doi.org/10.1007/978-3-031-29842-4

12. P. Ashwin, C. Perryman, and S. Wieczorek, *Parameter shifts for nonautonomous systems in low dimension: Bifurcation-and rate-induced tipping*, Nonlinearity **30** (2017), no. 6, 2185–2210.

13. P. Ashwin, S. Wieczorek, R. Vitolo, and P. Cox, *Tipping points in open systems: Bifurcation, noise-induced and rate-dependent examples in the climate system*, Phil. Trans. R. Soc. A **370** (2012), 1166–1184.

14. B. Aulbach, *A reduction principle for nonautonomous differential equations*, Archiv der Mathematik **39** (1982), 217–232.

15. B. Aulbach and J. Kalkbrenner, *Exponential forward splitting for noninvertible difference equations*, Comput. Math. Appl. **42** (2001), 743–754.

16. B. Aulbach and S. Siegmund, *The dichotomy spectrum for noninvertible systems of linear difference equations*, J. Difference Equ. Appl. **7** (2001), no. 6, 895–913.

17. ———, *A spectral theory for nonautonomous difference equations*, Proc. 5th Intern. Conference of Difference Eqns. and Applications (Temuco, Chile, 2000) (J. López-Fenner, et al, eds.), Taylor & Francis, London, 2002, pp. 45–55.

18. B. Aulbach and N. Van Minh, *The concept of spectral dichotomy for linear difference equations II*, J. Difference Equ. Appl. **2** (1996), 251–262.

19. B. Aulbach, N. Van Minh, and P.P. Zabreiko, *The concept of spectral dichotomy for linear difference equations*, J. Math. Anal. Appl. **185** (1994), 275–287.

20. B. Aulbach and T. Wanner, *Integral manifolds for Carathéodory type differential equations in Banach spaces*, Six Lectures on Dynamical Systems (B. Aulbach and F. Colonius, eds.), World Scientific, Singapore etc., 1996, pp. 45–119.

21. ———, *The Hartman–Grobman theorem for Carathéodory-type differential equations in Banach spaces*, Nonlin. Analysis (TMA) **40** (2000), 91–104.

22. A. Avila and J. Bochi, *A formula with some applications to the theory of Lyapunov exponents*, Israel J. Math. **131** (2002), 125–137.

23. D. Barkley, I.G. Kevrekidis, and A.M. Stuart, *The moment map: Nonlinear dynamics of density evolution via a few moments*, SIAM J. Appl. Dyn. Syst. **5** (2006), no. 3, 403–434.

24. L. Barreira and Y.P. Pesin, *Introduction to smooth ergodic theory*, Graduate Studies in Mathematics, vol. 148, AMS, Providence, RI, 2013.

25. A.G. Baskakov, *Invertibility and the Fredholm property of difference operators*, Mathematical Notes **67** (2000), no. 6, 690–698.

26. P.H. Baxendale, *A stochastic Hopf bifurcation*, Probab. Theory Related Fields **99** (1994), no. 4, 581–616.

27. A. Ben-Artzi and I. Gohberg, *Dichotomy, discrete Bohl exponents, and spectrum of block weighted shifts*, Integral Equations Oper. Theory **14** (1991), no. 5, 613–677.

28. ———, *Dichotomies of perturbed time varying systems and the power method*, Indiana Univ. Math. J. **42** (1993), no. 3, 699–720.

29. A. Berger, *Counting uniformly attracting solutions of nonautonomous differential equations*, Discrete Contin. Dyn. Syst. (Series S) **1** (2008), no. 1, 15–25.

30. A. Berger and S. Siegmund, *Uniformly attracting solutions of nonautonomous differential equations*, Nonlin. Analysis (TMA) **68** (2008), no. 12, 3789–3811.

31. W.-J. Beyn and J.-M. Kleinkauf, *The numerical computation of homoclinic orbits for maps*, SIAM J. Numer. Anal. **34** (1997), no. 3, 1209–1236.

32. W.-J. Beyn, G. Froyland, and T. Hüls, *Angular values of nonautonomous and random linear dynamical systems: Part I – Fundamentals*, SIAM J. Appl. Dyn. Syst. **21** (2022), no. 2, 1245–1286.

33. W.-J. Beyn and T. Hüls, *Angular values of linear dynamical systems and their connection to the dichotomy spectrum*, 2020, Preprint, arXiv:2012.11340.

34. Z. Bishnani and R.S. Mackey, *Safety criteria for aperiodically forced systems*, Dyn. Syst. **18** (2003), no. 2, 107–129.

35. C. Blachut and C. González-Tokman, *A tale of two vortices: How numerical ergodic theory and transfer operators reveal fundamental changes to coherent structures in non-autonomous dynamical systems*, J. Comput. Dyn. **7** (2020), no. 2, 369–399.

36. M. Bortolan, A. Carvalho, and J. Langa, *Attractors under autonomous and non-autonomous perturbations*, Mathematical Surveys and Monographs, vol. 246, AMS, Providence, RI, 2020.

37. R.T. Botts, A.J. Homburg, and T.R. Young, *The Hopf bifurcation with bounded noise*, Discrete Contin. Dyn. Syst. **32** (2012), no. 8, 2997–3007.

38. B. Braaksma, H.W. Broer, and G. Huitema, *Towards a quasi-periodic bifurcation theory*, Mem. Amer. Math. Soc. **83** (1990), 83–167.

39. B. Braaksma, G. Huitema, and F. Takens, *Unfoldings of quasi-periodic tori*, Mem. Amer. Math. Soc. **83** (1990), 1–82.

40. M. Callaway, T.S. Doan, J.S.W. Lamb, and M. Rasmussen, *The dichotomy spectrum for random dynamical systems and pitchfork bifurcations with additive noise*, Ann. Inst. Henri Poincaré Probab. Stat. **53** (2017), no. 4, 1548–1574.

41. S.L. Campbell and C.D. Meyer, *Generalized inverses of linear transformations*, Surveys and reference works in mathematics, vol. 4, Pitman, San Francisco, 1979.

42. T. Caraballo, J.A. Langa, R. Obaya, and A.M. Sanz, *Global and cocycle attractors for non-autonomous reaction-diffusion equations. The case of null upper Lyapunov exponent*, J. Differ. Equations **265** (2018), no. 9, 3914–3951.

43. A.N. Carvalho, J.A. Langa, and J.C. Robinson, *Attractors for infinite-dimensional non-autonomous dynamical systems*, Applied Mathematical Sciences, vol. 182, Springer, Berlin etc., 2012.

44. Á. Castañeda and G. Robledo, *Differentiability of Palmer's linearization theorem and converse result for density functions*, J. Differ. Equations **259** (2015), 4634–4650.

45. Á. Castañeda, G. Robledo, and P. Monzón, *Smoothness of topological equivalence on the half line for nonautonomous systems*, Proc. Roy. Soc. Edinb., Sect. A, Math. **150** (2020), no. 5, 2484–2502.

46. D.N. Cheban, P.E. Kloeden, and B. Schmalfuß, *The relationship between pullback, forward and global attractors of nonautonomous dynamical systems*, Nonlin. Dynam. Syst. Theory **2** (2002), 9–28.

47. A. Chenciner and G. Iooss, *Bifurcations de tores invariants*, Arch. Ration. Mech. Anal. **69** (1979), 108–198.

48. S.-N. Chow and J.K. Hale, *Methods of bifurcation theory*, Grundlehren der mathematischen Wissenschaften, vol. 251, Springer, Berlin etc., 1996.

49. S.-N. Chow, K. Lu, and Y.-Q. Shen, *Normal form and linearization for quasiperiodic systems*, Trans. Am. Math. Soc. **331** (1992), no. 1, 361–376.

50. C.V. Coffman and J.J. Schäffer, *Dichotomies for linear difference equations*, Math. Ann. **172** (1967), 139–166.

51. F. Colonius, R. Fabbri, R.A. Johnson, and M. Spadini, *Bifurcation phenomena in control flows*, Topol. Metholds Nonlinear Anal. **30** (2007), no. 1, 87–111.

52. F. Colonius and W. Kliemann, *The dynamics of control*, Birkhäuser, Basel etc., 1999.

53. ———, *Dynamical systems and linear algebra*, Graduate Studies in Mathematics, vol. 158, AMS, Providence, RI, 2014.

54. W.A. Coppel, *Dichotomies in stability theory*, Lect. Notes Math., vol. 629, Springer, Berlin etc., 1978.

55. H. Crauel and F. Flandoli, *Additive noise destroys a pitchfork bifurcation*, J. Dyn. Differ. Equations **10** (1998), no. 2, 259–274.

56. H. Crauel, P. Imkeller, and M. Steinkamp, *Bifurcations of one-dimensional stochastic differential equations*, Stochastic dynamics (H. Crauel and M. Gundlach, eds.), Springer, Berlin etc., 1999, pp. 27–47.

57. J.D. Crawford, *Introduction to bifurcation theory*, Rev. Mod. Phys. **63** (1991), 991–1037.

58. L.V. Cuong, T.S. Doan, and S. Siegmund, *A Sternberg theorem for nonautonomous differential equations*, J. Dyn. Differ. Equations **31** (2019), 1279–1299.

59. J.L. Daleckiĭ and M.G. Kreĭn, *Stability of solutions of differential equations in Banach space*, Translations of Mathematical Monographs, vol. 43, AMS, Providence, RI, 1974.

60. K. Deimling, *Nonlinear functional analysis*, Springer, Berlin etc., 1985.

61. T. Diagana, S. Elaydo, and A.-A. Yakubu, *Population models in almost periodic environments*, J. Difference Equ. Appl. **13** (2007), no. 4, 239–260.

62. L. Dieci, C. Elia, and E.S. van Vleck, *Exponential dichotomy on the real line: SVD and QR methods*, J. Differ. Equations **248** (2010), 287–308.

63. L. Dieci and E.S. van Vleck, *Lyapunov and other spectra: A survey*, Collected Lectures on the Preservation of Stability under Discretization, SIAM, 2002, pp. 197–218.

64. ———, *Lyapunov and Sacker–Sell spectral intervals*, J. Dyn. Differ. Equations **19** (2007), no. 2, 265–293.

65. C.G.H. Diks and F.O.O. Wagener, *A bifurcation theory for a class of discrete time Markovian stochastic systems*, Physica D **237** (2008), no. 24, 3297–3306.

66. T.S. Doan, M. Engel, J.S.W. Lamb, and M. Rasmussen, *Hopf bifurcation with additive noise*, Nonlinearity **31** (2018), no. 10, 4567–4601.

67. T.S. Doan, J.S.W. Lamb, J. Newman, and M. Rasmussen, *Classification of random circle homeomorphisms up to topological conjugacy*, 2020, Preprint, arXiv:1707.05401.

68. T.S. Doan, K.J. Palmer, and M. Rasmussen, *The Bohl spectrum for linear nonautonomous differential equations*, J. Dynam. Differential Equations **29** (2017), no. 4, 1459–1485.

69. T.S. Doan, M. Rasmussen, and P.E. Kloeden, *The mean-square dichotomy spectrum and a bifurcation to a mean-square attractor*, Discrete Contin. Dyn. Syst. Ser. B **20** (2015), no. 3, 875–887.

70. E.J. Doedel, R.C. Paenroth, A.R. Champneys, T.F. Fairgrieve, Y.A. Kuznetsov, B.E. Oldeman, B. Sandstede, and X. Wang, *AUTO 2000: Continuation and bifurcation software for ordinary differential equations (with HomCont)*, https://nlds.sdsu.edu/resources/auto2000.pdf, 2002.

71. J. Dueñas, C. Núñez, and R. Obaya, *Bifurcation theory of attractors and minimal sets in D-concave nonautonomous scalar ordinary differential equations*, 2022, Preprint, arXiv:2206.06728.

72. S. Elaydi, *An introduction to difference equations*, Undergraduate Texts in Mathematics, Springer, New York, 2005.

73. S. Elaydi and O. Hajek, *Exponential trichotomy of differential systems*, J. Math. Anal. Appl. **129** (1988), 362–374.

74. C. Elia and R. Fabbri, *Rotation number and exponential dichotomy for linear Hamiltonian systems: From theoretical to numerical results*, J. Dyn. Differ. Equations **25** (2013), 95–120.

75. M. Engel, J.S.W. Lamb, and M. Rasmussen, *Bifurcation analysis of a stochastically driven limit cycle*, Comm. Math. Phys. **365** (2019), no. 3, 935–942.

76. _____, *Conditioned Lyapunov exponents for random dynamical systems*, Trans. Amer. Math. Soc. **372** (2019), no. 9, 6343–6370.

77. R. Fabbri and R.A. Johnson, *On a saddle-node bifurcation in a problem of quasi-periodic harmonic forcing*, EQUADIFF 2003. Proceedings of the international conference on differential equations, Hasselt, Belgium, July 22–26, 2003 (Hackensack, NJ) (Dumortier, F., et al, eds.), World Scientific, 2005, pp. 839–847.

78. R. Fabbri, R.A. Johnson, and F. Mantellini, *A nonautonomous saddle-node bifurcation pattern*, Stoch. Dyn. **4** (2004), no. 3, 335–350.

79. A.M. Fink, *Almost periodic differential equations*, Lect. Notes Math., vol. 377, Springer, Berlin etc., 1974.

80. F. Flandoli, B. Gess, and M. Scheutzow, *Synchronization by noise*, Probab. Theory Related Fields **168** (2017), no. 3–4, 511–556.

81. _____, *Synchronization by noise for order-preserving random dynamical systems*, Ann. Probab. **45** (2017), no. 2, 1325–1350.

82. M. Franca and R.A. Johnson, *Remarks on nonautonomous bifurcation theory*, Rend. Istit. Mat. Univ. Trieste **49** (2017), 215–243.

83. M. Franca, R.A. Johnson, and V. Muñoz-Villarragut, *On the nonautonomous Hopf bifurcation problem*, Discrete Contin. Dyn. Syst. (Series S) **9** (2016), no. 4, 1119–1148.

84. G. Froyland, S. Lloyd, and A. Quas, *Coherent structures and isolated spectrum for Perron–Frobenius cocycles*, Ergodic Theory Dyn. Syst. **30** (2010), no. 3, 729–756.

85. G. Froyland and K. Padberg, *Almost-invariant sets and invariant manifolds – Connecting probabilistic and geometric descriptions of coherent structures in flows*, Physica D **238** (2009), 1507–1523.

86. G. Fuhrmann, *Non-smooth saddle-node bifurcations. I: Existence of an SNA*, Ergodic Theory Dyn. Syst. **36** (2016), no. 4, 1130–1155.

87. ———, *Non-smooth saddle-node bifurcations. III: Strange attractors in continuous time*, J. Differ. Equations **261** (2016), no. 3, 2109–2140.

88. G. Fuhrmann, M. Gröger, and T. Jäger, *Non-smooth saddle-node bifurcations. II: Dimensions of strange attractors*, Ergodic Theory Dyn. Syst. **38** (2018), no. 8, 2989–3011.

89. P. Giesl and M. Rasmussen, *A note on almost periodic variational equations*, Commun. Pure Appl. Anal. **10** (2011), no. 3, 983–994.

90. P. Glendinning, *Intermittency and strange nonchaotic attractors in quasiperiodically forced circle maps*, Physics Letters A **244** (1998), 545–550.

91. P. Glendinning, T. Jäger, and G. Keller, *How chaotic are strange non-chaotic attractors*, Nonlinearity **19** (2006), no. 9, 2005–2022.

92. W. Govaerts, Y.A. Kuznetsow, A. Yu, and B. Sijnava, *Bifurcation of maps in the software package CONTENT*, Proceedings of the 2nd Workshop on Computer Algebra in Scientific Computing. CASC '99 (Ganzha, V.G., et al, eds.), Springer, 1999, pp. 191–206.

93. C. Grebogi, E. Ott, S. Pelikan, and J.A. Yorke, *Strange attractors that are not chaotic*, Physica D **13** (1984), 261–268.

94. J. Guckenheimer and P.J. Holmes, *Nonlinear oscillations, dynamical systems, and bifurcations of vector fields*, Applied Mathematical Sciences, vol. 42, Springer, Berlin etc., 1983.

95. I. Győri and M. Pituk, *The converse of the theorem on stability by the first approximation for difference equations*, Nonlin. Analysis (TMA) **47** (2001), 4635–4640.

96. J.K. Hale, *Ordinary differential equations*, Robert E. Krieger Publishing Company, Huntington–New York, 1980.

97. G. Haller, *Finding finite-time invariant manifolds in two-dimensional velocity fields*, Chaos **10** (2000), no. 1, 99–108.

98. Y. Hamaya, *Bifurcation of almost periodic solutions in difference equations*, J. Difference Equ. Appl. **10** (2004), no. 3, 257–297.

99. D. Henry, *Geometric theory of semilinear parabolic equations*, Lect. Notes Math., vol. 840, Springer, Berlin etc., 1981.

100. M.R. Herman, *Une méthode pour minorer les exposants de Lyapunov et quelques exemples montrant le caractère local d'un théorème d'Arnold et de moser sur le tore de dimension 2*, Comm. Math. Helv. **58** (1983), 453–502.

101. D. Hinrichsen and A.J. Pritchard, *Mathematical systems theory I – Modelling, state space analysis, stability and robustness*, Texts in Applied Mathematics, vol. 48, Springer, Berlin etc., 2005.

102. M.W. Hirsch and S. Smale, *Differential equations, dynamical systems, and linear algebra*, Academic Press, Boston–New York–San Diego, 1974.

103. J.M. Holtzman, *Explicit ϵ and δ for the implicit function theorem*, SIAM Review **12** (1970), no. 2, 284–286.

104. A.J. Homburg and T.R. Young, *Bifurcations of random differential equations with bounded noise on surfaces*, Topol. Methods Nonlinear Anal. **35** (2010), no. 1, 77–97.

105. W. Horsthemke and R. Lefever, *Noise-induced transitions*, Springer Series in Synergetics, vol. 15, Springer, Berlin, 1984.

106. T. Hüls, *A model function for non-autonomous bifurcations of maps*, Discrete Contin. Dyn. Syst. (Series B) **7** (2007), no. 2, 351–363.

107. ———, *Numerical computation of dichotomy rates and projectors in discrete time*, Discrete Contin. Dyn. Syst. (Series B) **12** (2009), no. 1, 109–131.

108. ———, *Computing Sacker–Sell spectra in discrete time dynamical systems*, SIAM Journal on Numerical Analysis **48** (2010), no. 6, 2043–2064.

109. ———, *Computing stable hierarchies of fiber bundles*, Discrete Contin. Dyn. Syst. (Series B) **22** (2017), no. 9, 3341–3367.

110. T. Hüls and C. Pötzsche, *Qualitative analysis of a nonautonomous Beverton–Holt Ricker model*, SIAM J. Applied Dynamical Systems **13** (2014), no. 4, 1442–1488.

111. T. Hüls and Y.-K. Zou, *On computing heteroclinic trajectories of non-autonomous maps*, Discrete Contin. Dyn. Syst. (Series B) **17** (2012), no. 1, 79–99.

112. G. Iooss, *Bifurcation of maps and applications*, Mathematics Studies, vol. 36, North-Holland, Amsterdam etc., 1979.

113. M. Izydorek and S. Rybicki, *Bifurcation of bounded solutions of* 1-*parameter ODE's*, J. Differ. Equations **130** (1996), 267–276.

114. T. Jäger, *Elliptic stars in a chaotic night*, J. Lond. Math. Soc. **84** (2011), no. 3, 595–611.

115. A. Jänig, *Nonautonomous Conley index theory. Continuation of Morse-decompositions*, Topol. Metholds Nonlinear Anal. **53** (2019), no. 1, 79–96.

116. ———, *Nonautonomous Conley index theory. The connecting homomorphism*, Topol. Metholds Nonlinear Anal. **53** (2019), no. 2, 427–446.

117. ———, *Nonautonomous Conley index theory. The homology index and attractor-repeller decompositions*, Topol. Metholds Nonlinear Anal. **53** (2019), no. 1, 57–77.

118. L. Jiang, *Generalized exponential dichotomy and global linearization*, J. Math. Anal. Appl. **315** (2005), 474–490.

119. R.A. Johnson, *Exponential dichotomy, rotation number and linear differential operators with bounded coefficients*, J. Differ. Equations **61** (1986), 54–78.

120. ———, *Hopf bifurcation from nonperiodic solutions of differential equations. I. Linear theory*, J. Dyn. Differ. Equations **1** (1989), no. 2, 179–198.

121. R.A. Johnson, P.E. Kloeden, and R. Pavani, *Two-step transitions in nonautonomous bifurcations: An explanation*, Stoch. Dyn. **2** (2002), no. 1, 67–92.

122. R.A. Johnson and F. Mantellini, *A nonautonomous transcritical bifurcation problem with an application to quasi-periodic bubbles*, Discrete Contin. Dyn. Syst. **9** (2003), no. 1, 209–224.

123. R.A. Johnson, K.J. Palmer, and G.R. Sell, *Ergodic properties of linear dynamical systems*, SIAM J. Math. Anal. **18** (1987), no. 1, 1–33.

124. R.A. Johnson and Y. Yi, *Hopf bifurcation from nonperiodic solutions of differential equations. II*, J. Differ. Equations **107** (1994), no. 2, 310–340.

125. À. Jorba, F.J. Muñoz-Almaraz, and J.C. Tatjer, *On non-smooth pitchfork bifurcations in invertible quasi-periodically forced 1-d maps*, J. Difference Equ. Appl. **24** (2018), no. 4, 588–608.

126. N. Ju and S. Wiggins, *On roughness of exponential dichotomy*, J. Math. Anal. Appl. **262** (2001), 39–49.

127. N. Ju, D. Small, and S. Wiggins, *Existence and computation of hyperbolic trajectories of aperiodically time dependent vector fields and their approximations*, Int. J. Bifurcation Chaos **13** (2003), no. 6, 1449–1457.

128. T. Kato, *Perturbation theory for linear operators*, corrected 2nd ed., Grundlehren der mathematischen Wissenschaften, vol. 132, Springer, Berlin etc., 1980.

129. A. Katok and B. Hasselblatt, *Introduction to the modern theory of dynamical systems*, Encyclopedia of Mathematics and Its Applications, vol. 54, Cambridge Univ. Press, Cambridge, 1995.

130. G. Keller, *A note on strange nonchaotic attractors*, Fundamenta Mathematicae **151** (1996), no. 2, 139–148.

131. H. Keller and G. Ochs, *Numerical approximation of random attractors*, Stochastic dynamics (H. Crauel and M. Gundlach, eds.), Springer, Berlin etc., 1999, pp. 93–115.

132. H. Kielhöfer, *Bifurcation theory: An introduction with applications to PDEs*, second ed., Applied Mathematical Sciences, vol. 156, Springer, Berlin etc., 2012.

133. P.E. Kloeden, *Pullback attractors in nonautonomous difference equations*, J. Difference Equ. Appl. **6** (2000), no. 1, 33–52.

134. _____, *Pitchfork and transcritical bifurcations in systems with homogenous nonlinearities and an almost periodic time coefficient*, Commun. Pure Appl. Anal. **1** (2002), no. 4, 1–14.

135. P.E. Kloeden and P. Marín-Rubio, *Negatively invariant sets and entire solutions*, J. Dyn. Differ. Equations **23** (2011), no. 3, 437–450.

136. P.E. Kloeden and C. Pötzsche, *Nonautonomous bifurcation scenarios in SIR models*, Mathematical Methods in the Applied Sciences **38** (2015), 3495–3518.

137. P.E. Kloeden, C. Pötzsche, and M. Rasmussen, *Discrete-time nonautonomous dynamical systems*, Stability and Bifurcation Theory for Non-Autonomous Differential Equations (R.A. Johnson and M.P. Pera, eds.), Lect. Notes Math., vol. 2065, Springer, Berlin etc., 2012, pp. 35–102.

138. _____, *Limitations of pullback attractors for processes*, J. Difference Equ. Appl. **18** (2012), no. 4, 693–701.

139. P.E. Kloeden and M. Rasmussen, *Nonautonomous dynamical systems*, Mathematical Surveys and Monographs, vol. 176, AMS, Providence, RI, 2011.
140. P.E. Kloeden and S. Siegmund, *Bifurcations and continuous transitions of attractors in autonomous and nonautonomous systems*, Int. J. Bifurcation Chaos **5** (2005), no. 2, 1–21.
141. P.E. Kloeden and M. Yang, *An introduction to nonautonomous dynamical systems and their attractors*, Interdisciplinary Mathematical Sciences, vol. 21, World Scientific, Singapore etc., 2021.
142. M.A. Krasnosel'skij, V.Sh. Burd, and Yu.S. Kolesov, *Nonlinear almost periodic oscillations*, A Halsted Press Book, John Wiley & Sons, Jerusalem–London, 1973.
143. B. Krauskopf, H.M. Osinga, and J. Galán-Vioque (eds.), *Numerical continuation methods for dynamical systems. Path following and boundary value problems*, Understanding Complex Systems. Springer, Dordrecht, 2007.
144. S.G. Kryzhevich and V.A. Pliss, *Structural stability of nonautonomous systems*, Differential Equations **39** (2003), no. 10, 1395–1403.
145. C. Kuehn, G. Malavolta, and M. Rasmussen, *Early-warning signals for bifurcations in random dynamical systems with bounded noise*, J. Math. Anal. Appl. **464** (2018), no. 1, 58–77.
146. J. Kurzweil, *Ordinary differential equations*, Studies in Applied Mathematics 13, Elsevier, Amsterdam etc., 1986.
147. N.V. Kuznetsov, *Stability and oscillations of dynamical systems – Theory and applications*, Jyväskylä Studies in Computing, vol. 96, Birkhäuser, Jyväskylä, 2008.
148. Y.A. Kuznetsov, *Elements of applied bifurcation theory*, 3rd ed., Applied Mathematical Sciences, vol. 112, Springer, Berlin etc., 2004.
149. J.S.W. Lamb, M. Rasmussen, and C.S. Rodrigues, *Topological bifurcations of minimal invariant sets for set-valued dynamical systems*, Proc. Am. Math. Soc. **143** (2015), no. 9, 3927–3937.
150. J.A. Langa, J.C. Robinson, and A. Suárez, *Stability, instability, and bifurcation phenomena in non-autonomous differential equations*, Nonlinearity **15** (2002), 887–903.
151. _____, *Bifurcation from zero of a complete trajectory for nonautonomous logistic PDEs*, Int. J. Bifurcation Chaos **15** (2005), no. 8, 2663–2669.
152. _____, *Bifurcations in non-autonomous scalar equations*, J. Differ. Equations **221** (2006), 1–35.
153. G.A. Leonov, *Strange attractors and classical stability theory*, University Press, St. Petersburg, 2008.
154. G.A. Leonov and N.V. Kuznetsov, *Time-varying linearization and the Perron effects*, Int. J. Bifurcation Chaos Appl. Sci. Eng. **17** (2007), no. 4, 1079–1107.
155. L.M. Lerman and E.V. Gubina, *Nonautonomous gradient-like vector fields on the circle: Classification, structural stability and autonomization*, Discrete Contin. Dyn. Syst. (Series S) **13** (2020), no. 4, 1341–1367.
156. L.M. Lerman and L.P. Shil'nikov, *On the classification of structurally stable nonautonomous systems of second order with a finite number of cells*, Dokl.

Akad. Nauk SSSR (also Soviet Math. Dokl.) **Tom 209 (resp. 14)** (1973), no. 3 (resp. 2), 444–448.

157. _____, *Homoclinical structures in nonautonomous systems: Nonautonomous chaos*, Chaos **2** (1992), no. 3, 447–454.

158. Ta Li, *Die Stabilitätsfrage bei Differenzengleichungen*, Acta Math. **63** (1934), 99–141.

159. K.K. Lin and L.-S. Young, *Shear-induced chaos*, Nonlinearity **21** (2008), no. 5, 899–922.

160. I.P. Longo, C. Núñez, and R. Obaya, *Critical transitions in piecewise uniformly continuous concave quadratic ordinary differential equations*, arXiv:2110.10145.

161. I.P. Longo, C. Núñez, R. Obaya, and M. Rasmussen, *Rate-induced tipping and saddle-node bifurcation for quadratic differential equations with nonautonomous asymptotic dynamics*, SIAM J. Appl. Dyn. Syst. **20** (2021), no. 1, 500–540.

162. A.M. Lyapunov, *The general problem of the stability of motion*, Mathematical Society of Kharkov, Kharkov, 1892, (in Russian).

163. T. Ma and S. Wang, *Bifurcation theory and applications*, Series on Nonlinear Sciences, vol. 53, World Scientific, Hackensack, NJ, 2005.

164. R.M. May, *Simple mathematical models with very complicated dynamics*, Nature **261** (1976), 459–467.

165. J. Milnor, *On the concept of attractor*, Comm. Math. Phys. **99** (1985), 177–195.

166. M. Ndour, K. Padberg-Gehle, and M. Rasmussen, *Spectral warning signs for sudden changes in time-dependent flow patterns*, Fluids **2021**, 6(2), 49.

167. S.S. Negi, A. Prasad, and R. Ramaswamy, *Bifurcations and transitions in the quasiperiodically driven logistic map*, Physica D **1–2** (2000), no. 145, 1–12.

168. _____, *Strange nonchaotic attractors*, Int. J. Bifurcation Chaos **2** (2001), no. 11, 291–309.

169. V.V. Nemytskii and V.V. Stepanov, *Qualitative theory of differential equations*, University Press, Princeton, NJ, 1960.

170. T.Y. Nguyen, T.S. Doan, T. Jäger, and S. Siegmund, *Non-autonomous saddle-node bifurcations in the quasiperiodically forced logistic map*, Int. J. Bifurcation Chaos **21** (2011), no. 5, 1427–1438.

171. S. Novo, C. Núñez, and R. Obaya, *Ergodic properties and rotation number for linear Hamiltonian systems*, J. Differ. Equations **148** (1998), no. 1, 148–185.

172. S. Novo, R. Obaya, and A.M. Sanz, *Almost periodic and almost automorphic dynamics for scalar convex differential equations*, Isr. J. Math. **144** (2004), 157–189.

173. C. Núñez and R. Obaya, *A non-autonomous bifurcation theory for deterministic scalar differential equations*, Discrete Contin. Dyn. Syst. (Series B) **9** (2008), no. 3–4, 701–730.

174. _____, *Li–Yorke chaos in nonautonomous Hopf bifurcation patterns. I.*, Nonlinearity **32** (2019), no. 10, 3940–3980.

175. R. Obaya, and A.M. Sanz, *Non-autonomous scalar linear-dissipative and purely dissipative parabolic PDEs over a compact base flow*, J. Differ. Equations **285** (2021), 714–750.

176. E. Ott, *Strange attractors and chaotic motions of dynamical systems*, Rev. Mod. Phys. **53** (1981), 655–671.

177. K.J. Palmer, *A generalization of Hartman's linearization theorem*, J. Math. Anal. Appl. **41** (1973), 753–758.

178. _____, *The structurally stable linear systems on the half-line are those with an exponential dichotomy*, J. Differ. Equations **33** (1979), 16–25.

179. _____, *Exponential dichotomies, the shadowing lemma and transversal homoclinic points*, Dynamics Reported (U. Kirchgraber and H.-O. Walther, eds.), vol. 1, B.G. Teubner/John Wiley & Sons, Stuttgart/Chichester etc., 1988, pp. 265–306.

180. G. Papaschinopoulos, *Exponential dichotomy for almost periodic linear difference equations*, Ann. Soc. Sci. Bruxelles, Sér. I **102** (1988), no. 1–2, 19–28.

181. _____, *On exponential trichotomy of linear difference equations*, Appl. Anal. **40** (1991), 89–109.

182. J. Pejsachowicz and R. Skiba, *Topology and homoclinic trajectories of discrete dynamical systems*, Discrete Contin. Dyn. Syst. (Series S) **6** (2013), no. 4, 1077–1094.

183. O. Perron, *Die Stabilitätsfrage bei Differentialgleichungen*, Math. Z. **32** (1930), 703–728.

184. H. Poincaré, *L'équilibre d'une masse fluide animée d'un mouvement de rotation*, Acta Math. **7** (1885), no. 259–380.

185. C. Pötzsche, *A note on the dichotomy spectrum*, J. Difference Equ. Appl. **15** (2009), no. 10, 1021–1025 (corrigendum in J. Difference Equ. Appl. **18** (2012), no. 7, 1257–1261).

186. _____, *Geometric theory of discrete nonautonomous dynamical systems*, Lect. Notes Math., vol. 2002, Springer, Berlin etc., 2010.

187. _____, *Nonautonomous bifurcation of bounded solutions I: A Lyapunov–Schmidt approach*, Discrete Contin. Dyn. Syst. (Series B) **14** (2010), no. 2, 739–776.

188. _____, *Bifurcations in nonautonomous dynamical systems: Results and tools in discrete time*, Proceedings of the workshop on future directions in difference equations, Vigo, Spain, June 13–17, 2011 (Vigo) (E. Liz, ed.), Colección on Congresos, no. 69, Servizo de Publicacións de Universidade de Vigo, 2011, pp. 163–212.

189. _____, *Nonautonomous bifurcation of bounded solutions II: A shovel bifurcation pattern*, Discrete Contin. Dyn. Syst. (Series A) **31** (2011), no. 1, 941–973.

190. _____, *Nonautonomous continuation of bounded solutions*, Commun. Pure Appl. Anal. **10** (2011), no. 3, 937–961.

191. _____, *Persistence and imperfection of nonautonomous bifurcation patterns*, J. Differ. Equations **250** (2011), 3874–3906.

192. _____, *Corrigendum to "A note on the dichotomy spectrum"*, J. Difference Equ. Appl. **18** (2012), no. 7, 1257–1261.

193. _____, *Nonautonomous bifurcation of bounded solutions: Crossing curve situations*, Stoch. Dyn. **12** (2012), no. 2.

194. C. Pötzsche and M. Rasmussen, *Taylor approximation of invariant fiber bundles for nonautonomous difference equations*, Nonlin. Analysis (TMA) **60** (2005), no. 7, 1303–1330.

195. _____, *Taylor approximation of integral manifolds*, J. Dyn. Differ. Equations **18** (2006), no. 2, 427–460.

196. C. Pötzsche and R. Skiba, *A continuation principle for Fredholm maps II: Application to homoclinic solutions*, Math. Nachr. **293** (2020), no. 6, 1174–1199.

197. A. Prasad, V. Mehra, and R. Ramaswamy, *Strange nonchaotic attractors in the quasiperiodically forced logistic map.*, Physical Review E **2** (1998), no. 57, 1576–1584.

198. M. Rasmussen, *Towards a bifurcation theory for nonautonomous difference equations*, J. Difference Equ. Appl. **12** (2006), no. 3–4, 297–312.

199. _____, *Attractivity and bifurcation for nonautonomous dynamical systems*, Lect. Notes Math., vol. 1907, Springer, Berlin etc., 2007.

200. _____, *Nonautonomous bifurcation patterns for one-dimensional differential equations*, J. Differ. Equations **234** (2007), 267–288.

201. _____, *Dichotomy spectra and Morse decompositions of linear nonautonomous differential equations*, J. Differ. Equations **246** (2009), no. 6, 2242–2263.

202. _____, *An alternative approach to Sacker–Sell spectral theory*, J. Difference Equ. Appl. **16** (2010), no. 2–3, 227–242.

203. F. Remo, G. Fuhrmann, and T. Jäger, *On the effect of forcing of fold bifurcations and early-warning signals in population dynamics*, 2022, Preprint, arXiv:1904.06507.

204. P. Ritchie and J. Sieber, *Early-warning indicators for rate-induced tipping*, Chaos **26(9)** (2016), 093116, 13.

205. D. Ruelle, *Ergodic theory of differentiable dynamical systems*, Inst. Hautes Études Sci. Publ. Math. (1979), no. 50, 27–58.

206. _____, *Characteristic exponents and invariant manifolds in Hilbert space*, Ann. of Math. (2) **115** (1982), no. 2, 243–290.

207. E. Ruß, *Dichotomy spectrum for difference equations in Banach spaces*, J. Difference Equ. Appl. **23(3)** (2016), 576–617.

208. R.J. Sacker, *Bifurcation in the almost periodic Ricker map*, J. Difference Equ. Appl. **25** (2019), no. 5, 599–618.

209. _____, *A new metric yielding a richer class of unbounded functions having compact hulls in the shift flow*, J. Dyn. Differ. Equations **33** (2021), 833–848.

210. R.J. Sacker and G.R. Sell, *A spectral theory for linear differential systems*, J. Differ. Equations **27** (1978), 320–358.

211. K.R. Schenk-Hoppé, *Stochastic Hopf bifurcation: An example*, Int. J. Nonlin. Mech. **31** (1996), no. 5, 685–692.

212. M. Scheutzow and I. Vorkastner, *Synchronization, Lyapunov exponents and stable manifolds for random dynamical systems*, Stochastic Partial Differential Equations and Related Fields (A. Eberle, M. Grothaus, W. Hoh, M. Kassmann, W. Stannat, and G. Trutnau, eds.), Proceedings in Mathematics & Statistics, Springer, 2018, pp. 359–366.

213. G.R. Sell, *Topological dynamics and ordinary differential equations*, Van Nostrand Reinhold Mathematical Studies, no. 33, Van Nostrand Reinhold, London etc., 1971.

214. _____, *Bifurcation of higher dimensional tori*, Arch. Ration. Mech. Anal. **69** (1979), 199–230.

215. S. Siegmund, *Dichotomy spectrum for nonautonomous differential equations*, J. Dyn. Differ. Equations **14** (2002), no. 1, 243–258.

216. _____, *Normal forms for nonautonomous differential equations*, J. Differ. Equations **178** (2002), no. 2, 541–573.

217. _____, *Normal forms for nonautonomous difference equations*, Comput. Math. Appl. **45** (2003), no. 6–9, 1059–1073.

218. C.E. Silva, *Invitation to ergodic theory*, Student Mathematical Library, vol. 42, AMS, Providence, RI, 2007.

219. R. Skiba and N. Waterstraat, *The index bundle and multiparameter bifurcations for discrete dynamical systems*, Discrete Contin. Dyn. Syst. (Series A) **37** (2017), no. 11, 5603–5629.

220. N. Sri Namachchivaya, *Stochastic bifurcation*, Appl. Math. Comput. **38** (1990), no. 2, 101–159.

221. J. Stark, *Transitive sets for quasi-periodically forced monotone maps*, Dyn. Syst. **18** (2003), no. 4, 351–364.

222. J. Stark and R. Sturman, *Semi-uniform ergodic theorems and applications to forced systems*, Nonlinearity **13** (2000), no. 2, 113–143.

223. M. Steinkamp, *Bifurcations of one-dimensional stochastic differential equations*, Logos Verlag, Berlin, 2000.

224. V.I. Tkachenko, *On the exponential dichotomy of linear difference equations*, Ukr. Math. J. **48** (1996), no. 10, 1600–1608.

225. Q. Wang and L.-S. Young, *Strange attractors in periodically-kicked limit cycles and Hopf bifurcations*, Comm. Math. Phys. **240** (2003), no. 3, 509–529.

226. J.R. Ward, *Bifurcation of bounded solutions of ordinary differential equations*, J. Korean Math. Soc. **37(5)** (2000), 707–720.

227. S. Wieczorek, *Stochastic bifurcation in noise-driven lasers and Hopf oscillators*, Physical Review E (2009), no. 79, 1–10.

228. S. Wiggins, *Introduction to applied nonlinear dynamical systems and chaos*, Texts in Applied Mathematics, vol. 2, Springer, Berlin etc., 1990.

229. E. Zeidler, *Nonlinear functional analysis and its applications I (Fixed-points theorems)*, Springer, Berlin etc., 1993.

230. H. Zmarrou and A.J. Homburg, *Bifurcations of stationary measures of random diffeomorphisms*, Ergodic Theory Dyn. Syst. **27** (2007), no. 5, 1651–1692.

231. _____, *Dynamics and bifurcations of random circle diffeomorphisms*, Discrete Contin. Dyn. Syst. Ser. B **10** (2008), no. 2–3, 19–731.

Index

© The Author(s), under exclusive license to Springer Nature Switzerland AG 2023   153
V. Anagnostopoulou et al., *Nonautonomous Bifurcation Theory*, Frontiers
in Applied Dynamical Systems: Reviews and Tutorials 10,
https://doi.org/10.1007/978-3-031-29842-4

Printed in the United States
by Baker & Taylor Publisher Services